FOOS'S THEOREM OF G

The Lost Laws of Gravity Derived, Revised And Expanded

Al Foos

Montana State University

Foos's Theorem Of G - Escape From A Black Hole

Deriving Newton's G From Ink In A Bathtub
March, 2023 by Al Foos

The brainwashing from media hype and political support pushing Einstein's fraud wasn't fully appreciated until the slightly different interpretations of Einstein's theories were made manifest from dozens of posts in a physics Google group. Long ago I learned that questioning Einstein in a forum would result in immediate banishment on grounds of conspiracy theory. My posts were deleted by the Google group moderator, too, but only after being able to get a firm grasp of variations on the Einstein cult fostered by government. You can glimpse this yourself in Appendices A and B. I received dozens of threatening emails and phone calls, and the wiki article on Pound-Rebka was overwritten to conceal the obvious fraud addressed in my own posts. So, there you have it, proof the entire field of cosmology consists of government sponsored scientific fraud. The current cosmology model is a black hole filled with fake assumptions and math abuse, not a real spatially expanding universe that propagates light at an internal fixed rate of c:

The root problem is inherited from misrepresentation of the Pound-Rebka results claiming a variable c over a potential energy gradient. This originates way back to Newton over 300 years ago, who obviously couldn't have derived the laws of gravitation when he didn't understand the cause of gravity. His misguided corpuscular theory described in his famous Opticks published in 1700 reflected this fundamental error in the understanding of gravity and the properties of light. Huygens had publicly refuted Newton's theory by then, but it was tragically resurrected by Einstein in his general fake theory of relativity which claimed to best Newton's laws which it definitely did not. So, to shake off the burden of exposing Einstein to find the truth, it is necessary to sit down and derive the laws of gravity from the ground up here and now. This clears our minds of the brazen errors in algebra Einstein exploited to make fools of us masses so that we can have clear understanding of how spatial expansion works. A ninth grade algebra class with perfect marks should be enough to follow the math and verify that this Foos Theorem of G is perfectly correct. Few adults have attained or retained that proficiency, but with effort I guarantee the light will dawn on any devoted student of math and science:

So, this is indeed the Holy Grail of physics. Few are able to drink from it; indeed, only a handful know the truth. This is not a theory, but an elegant mathematical theorem that good scientists will recognize as one of the most important in history, if any can be found. They will easily affirm my claims and marvel at the farcical incompetence entrenched in modern physics and cosmology. The field is crammed with Einstein cultists who feed from the government trough. A government job is not a real challenge. They're role is to deride anyone who shares a doubt. They are vicious piranha with very poor math skills despite a grandiose pretense. Frauds like their daddy. The origin of this tragic corruption of science actually stems from Newton's incomplete version of the laws of gravity from a misunderstanding of its cause. His laws were geometrically perfect, but he had no idea of the cause of gravity. So here we sit 300 years later more ignorant than ever. The chore for me is to derive these laws from the ground up and communicate that understanding. It comes easy to me thanks to well polished math proficiency and a strong background in physical chemistry. This isn't about theory. It's like thousands of math and science problems that have entirely black and white answers, not theories. Conscience demands the truth be conveyed as the geometrically perfect description of reality that it is. Welcome to Foos's Theorem Of G. How can the truth of gravity and the universe be brought to the understanding of men who have been vigorously brainwashed from birth? That's why I use an informal, back door approach that leads to the bleak brink of the Einstein and Hawking black holes. First, the theorem is developed, then a back door method used to contrast the truth with the confusion of modern cosmology so you see it clearly for yourself. I wouldn't waste your time with less.

Whatever model of gravity used, the value of g at the surface of a universe can only be c/s because R and M are independent variables and the velocity of light c the dependent one. Oh, how those little Einsteins are screaming already. "Nutjob! Crackpot! g at radius R of the universe is not c/s. You delusional fool!" So, I'll have to do this job once. This is where it cuts deep, kids. If a universe has no mass, c=0. Stay on track. If it has mass = one, then c=1, and if it has mass = 2 then c=2 and so on. The speed of light depends on mass M. Your own formula that you claim is proved by Pound-Rebka says that the change in c is proportional to g. Little g equals MG/R^2. If the universe had twice the mass, c would be twice as fast, and objects would fall twice as fast even by your own twisted reasoning. That is right, kids, isn't it? But how is Einstein's original claim that photons fall twice as fast as ordinary objects sensible? Do they have fins? How did an idiot become the greatest scientist of all time? Let's derive Newton's G for ourselves and see how fast imaginary photons are or rather, how fast electromagnetic waves are propagated.

Method 1: Derivation of G by Foos from the ground up. Use your own head to affirm the proof of the theorem starting with the only reasonable cause of gravity and a formula for G. We don't need Newton's formula at all, do we? If we plop down some mass in the middle of nowhere, what do you think it's gonna do? Is it going to sit there, shrink or expand? What does a bottle of ink do when you toss it in the tub? If it sits there, is its gravity going to attract other objects at potential energy mgh? No. It will diffuse outwardly with force because that is the way thermodynamic potential differences behave. The ink won't gather itself up and ask to be put back in the bottle. Einstein's master's thesis was on Brownian motion, so he should have easily grasped the fact that matter will expand into a void, but he wasn't a scientist. Figures lie and liars figure. Unlike the diffusion of ink which slows until equilibrium is reached, matter diffusing into a void experiences no resistance and continues expanding at the same accelerated rate for eternity, so universal expansion manifested as gravity is the solitary exception to the second law of thermodynamics and resolves the paradox of how things got ordered in the first place. Matter being infinitely divisible in a true vacuum and with nothing to resist it, the equal and opposite force of (accelerating) expansion manifests as gravity at the rate c/s. Interstitial pores between clumps of matter fill with light propagated at the rate of expansion c/s proportional to mass M for the universe as a whole and at the rate of relative spatial expansion seen as redshift between centers of mass within.

Despite accelerated expansion, the rate would diminish with the square of radius R, the radius of an expanding circle. This is a geometric principle already well known from grade school, and therefore c/s would be proportional to M/R^2 as a geometric fact and immutable constant. Easy stuff. These expanding pores truly do consist of nothing within an infinite void, particles of ink would expand at rates proportional to their mass from point to point. The universe would remain internally fixed in proportion eternally. Is the light beginning to dawn? Elementary algebra tells us the maximum rate of expansion at the outermost rim is $c/s = \dfrac{KM}{R^2}$ where K is a constant of proportionality. This equation is just plain common sense. Does it look familiar? This is the root cause and law of gravity. Newton got the formula right, but not the cause and complete picture.

What if we move closer to the universe center and stop at distance R'? Then $\dfrac{\Delta c}{s} = -\Delta g = \dfrac{KM}{R_1^2} - \dfrac{KM}{R_0^2}$. Do you recognize the similarity to Newton's G and the Pound-Rebka expressions for fractional frequency? Instead of h, we use the correct delta notation for delta R learned in ninth grade. If the Einsteins did that they could tell up from down and their Asberger's from a black hole. We want the change in the rate of falling bodies over distance Ro to R1. The correct sign for delta R is negative, so you know light at constant c including redshift is accelerated away from mass and exerting the equal and opposite force of gravity by the third law of motion. Distance between points inside the box is proportionally constant; therefore net speed of light away from mass internally is a fixed c instead of the "outside the box" rate c/s.

What does delta R over c mean? Velocity c is distance/time, so (delta R)/c is the time t seconds it takes to get from R down to R prime. Multiplying both sides of an equation by the same thing is still an equality, remember? So, the fractional change in rate of expansion is $\dfrac{\Delta c}{c} = \dfrac{-\Delta g}{c} = \dfrac{\dfrac{KM}{R_1^2} - \dfrac{KM}{R_o^2}}{c}$

This is the same fractional change in frequency and/or c found in most interpretations of Pound-Rebka, except that gh is replaced with delta g so the correct result is negative. Light is forced away from mass at the same rate bodies fall in the opposite direction. We just derived this without the knowledge of either Newton or Einstein. If only Newton and Einstein could have had Mr. Bobinski for math and then you wouldn't be stuck in a black hole. It's likely confusing that c as rate of expansion progressively decreases (closer to the universe center) in the positive while (delta c)/s as g progressively increases in the negative at the same progressive rate and also the same rate that bodies fall. It's like crosstown traffic. The speed of light c with redshift accelerates away from Earth as spatial expansion while the rate of increase progressively declines with R; hence,

$\dfrac{\Delta c}{s} = \dfrac{KM}{R^2}$ if R'<R, and $\dfrac{c}{s} = -g = \dfrac{KM}{R^2}$. You can easily lose your bearings, at least

sometimes I do, and I suspect not many are proficient enough at juggling algebra or the world would have fixed this mess 300 years ago. A good way to keep your bearings is by pondering Figure 1: By now it should be agreeable that K and Newton's G are one and same, but now G and the force of gravity have a clear and far richer meaning.

I highly recommend changing your delta style and not confuse yourselves with c and c' and by remembering c was your initial point and c' your final point and consistently calculate delta c as c'-c. Review standard math procedures and be sure delta c is always final minus initial. A better choice of subscripts c(2) and c(1) or even c(f) and c(i) (for final and initial) might help keep your head on straight. If your dependent variable is e', then do not use g but delta g as gf-gi without h because the proportion is change in velocity not potential energy with velocity squared. Use the basic rules of math and get the Pound-Rebka formulas right and not backwards like the little Einsteins in Appendix B. From now on we use G instead of K, but remember how we simply inferred these relationships from the analogy of a bottle of ink in a bathtub. Newton could see how bodies fell, but the outward expansion of space as a cause wasn't so obvious.

Remember that all changes in c, usually shown as c', are not directly measurable even though apparently Einsteins and ether entrainment theorists believe they are and they gave me a real hard time on the Google group. The value of measured c is strictly invariant as Michelson-Morley and Louis Essen established seventy to 150 years ago. Even Einstein himself appears to have reversed his famous mind going from the special theory to the general theory, but abandon confusion and seek the truth. Gravitational changes in c, unlike the effect of glass, are strictly the result of change in meter length that shows up as redshift or blueshift. Whenever space expands, the value of c changes externally only and shows up inside the box as redshift. If you followed this derivation closely, you should be able to see this clearly and from now on abandon that notion of a variable c except for the effect redshift or blueshift has on unit length.

Method 2: The general physics community cites the Pound-Rebka results to cite their firm belief that c is variable with mass according to 1+gh/c^2 where gh/c^2 is fractional change in potential energy, PE, according to Einstein. Light falls literally at the same or twice the rate as objects of mass in this view reflected in the article by Valev in Appendix B: He regards Einstein a liar for coming up with c=twice that expression instead of one, maintaining the Pound-Rebka formula proves photons fall at the same rate, not twice. He cites numerous scientists sharing these two conflicting ideas except for Stephen Hawking who cites c as invariant because he didn't want to look stupid. It is a widely accepted fact that the invariant speed of light was established by Michelson-Morley in the 1800s and Louis Essen in 1949 and confirmed numerous times. Einstein himself became famous mostly as a result of the false impression that his special theory proved c was invariant. The great unwashed is yet firmly convinced he proved the earlier Michelson-Morley experiment of an invariant c. Most are unaware that in his general theory the speed of light decided to change like falling golf balls. Someone tell YouTube.

The basis of this variable c the majority of the scientific community believes is clearly the formula given for fractional change in potential energy used in the Pound-Rebka experiment we see in the wiki article (Appendix A). However, never was the speed of light any objective of the experiment, but only fractional frequency change which is wrongly claimed to increase proportional to PE as $1+gh/c^2$ and clock speed which would decrease according to $1-gh/c^2$. This assumes change in velocity equals change in PE per Einstein; however, it does not. The units on either side do cancel, but change in velocity of the falling crystal is not equal to change in potential energy. The numerators would be change in velocity equals change in velocity squared. The use of PE is another Rocky Mountain Einstein major fallacy which should be obvious to any proficient mathematician. Study the equations in Figure 1. Delta c is equal to delta c, duh, no surprise because measured c is always fixed. And fractional delta length, delta wavelength, and delta clock speed are all equal to delta g over c. Frequency is a special case because it increases at lower elevations; therefore,

delta f/f(initial) = $(GM/R1^2 - GM/Ro^2)/c$ instead of $(GM/R1^2 - GM/Ro^2)/c$

where $c = GM/R^2$ because in this special case initial g is at the surface. For all other cases between objects, such as in the Pound-Rebka experiment, the correct formula for fractional change in frequency is the same as the fractional change in g:

$$\frac{\Delta f}{f_i} = -\frac{\frac{GM}{R_f^2} - \frac{GM}{R_i^2}}{\frac{GM}{R_i^2}}, \text{ where i is initial state and f is final } (R_i > R_f).$$

Anyway, the experiment wasn't meant to measure a variable c or the Doppler shift of the falling crystal. The Highway Patrol easily does that. The scientific community's modern consensus that photons fall back at rate gh/c^2 after ejected at rate c stems from the erroneous assumption that photons (corpuscles) have mass or are affected as if they did. The Pound-Rebka experiment was never meant to decide that, only that the frequency of light itself is greater at lower elevations without need for photons, and this was supposed by inference to prove a slower clock speed and thus Einstein's general relativity, not a variable photon. If you don't have a good mind for problem solving, you just believe what nonsense you're told; otherwise, you can sort it out as I've done here.

To say I was shocked to discover this sharp contradiction in how the speed of light varies or not is an understatement. I didn't know Einstein had changed his mind with GR. The experiment was meant to determine gravitational redshift of light itself at different elevations, not the Doppler shift of falling photons. Or was it? It certainly didn't change the public's conviction that c was invariant which is still the subject of YouTube videos praising Einstein's eternal genius. Why clock speed? Note how the wiki article avoids a good explanation of that but promotes numerous equations of Einstein's that are easily shown false. But let's pursue this notion of clock speed.

The key is knowing what is meant by clock speed and why decrease in wavelength at lower elevations translates to decrease in clock speed. The real answer has nothing to do with Einstein but is based squarely on the accepted scientific definition of a meter stated in terms of either frequency (or wavelength) or the time it takes to get from point A to B:

If the meter is half as long due to half the frequency due to redshift, it takes half as long to get from end to end by the reckoning of a fixed observer. At elevation h a meter consists of n wavelengths of a spectral line of cesium. At elevation zero, n number of wavelengths is now greater, for the sake of argument, let's say 10n. Hence, at h, if it's assumed it takes one second to get from A to B, then at ground zero it would take only one tenth of a second because the relative distance covered is only 1/10th the upper distance between A and B. If the distance traveled at ground zero is only 1/10 of the distance covered at h, then the velocity of c is less at ground zero if the shorter wavelength is taken into account. These little Einstein's are badly confused. By strict definition of a meter, and not my idea or any theory, the velocity of c is slower at lower elevations given the length of a meter is shorter, but the measured value of c is constant. The original intent and true meaning of the experiment have been progressively garbled because of Einstein's fraudulent tinkering.

However, both lower and upper meters are unchanged by their own reckoning (local observers). At h, n wavelengths over a time t is a meter, and at ground zero, a meter is 10 wavelengths over time t/10, so the meters are locally unchanged even while relative to an observer at another elevation or at the center of a black hole. ***The unit of length itself varies by the observation of a fixed, remote***, not local, observer while the measured speed of c is constant. This is what the measurements of c by Michelson-Morley and Louis Essen confirm. The measured speed of light does not vary in any direction irrespective of elevation. Differences in the units themselves occur due to gravitational frequency shift. A wave length in the upper meter is longer, the lost point of the experiment, so the relative speed of c including redshift is greater to that same degree, but no difference in c is measured. These facts have nothing to do with theory, but by the definition of units which are the true fundamentals of science; hence, the equations derived by Foos in Figures 1 and 2 in The Big Bang Boozle (overlooking the use of potential energy gh). In this theorem we derive the real equations (see Figure 1) as geometric constants.

The positive sign in the expression These facts have nothing to do with theory, but by the definition of a meter, the most fundamental fact of science; hence, the equations derived by Foos in Figures 1 and 2 in The Big Bang Boozle and this theorem are accurate and their logical interpretation sound. They have nothing to do with theory, but by the definition of a meter; hence, the equations derived by Foos in Figures 1 and 2 in The Big Bang Boozle and this theorem are accurate and their logical interpretation sound. A change in speed of light from $c'=c(1+gh/c^2)$ is also a faulty interpretation of Pound-Rebka never intended by the experimenters to start with. The gh term also fails unit analysis. The units cancel, yes, but gh/c^2 is in the form of velocity squared; whereas the left side is velocity *not* squared. The measured speed of c was also long ago proved invariant. It may be that c can be considered variable, but only in terms of the degree to which gravitational frequency shift dictates how much space between points A and B have expanded or contracted according to $c'=c+\Delta g$, not $c'=1-gh/c^2$ (see Figure 1). If c is variable, light accelerates upward as redshift. There is no basis whatsoever for believing that there is such a thing as photons attracted to mass like golf balls despite Newton's theory of light and the botched version of the same by Einstein. That should be the end of that argument.

But we have not yet properly phrased and proved the theorem, so let's move on to how the theorem refutes the standard model and gives a logically consistent and experimentally supported cause of gravity. To help illustrate, the absurdity of a variable speed of light c is compared. Hawking's invariant c results in the same absurdity. He fails to recognize that in terms of redshift and the definition of light, imaginary photons are not trapped by gravity, but waves accelerate away from mass as redshift. The constancy of c and definition of a meter do not allow a different interpretation, but does that matter when cosmology is really only science fiction?

Proof: Foos's Theorem Of G

$G = R^2/M * (c/s)$ where R is radius of universe and M is mass of universe
$G = R^2/M * (c/s)$ where c is universal invariant speed of light
$G = R^2/M * (c/s)$ where $c/s = g$ = rate of falling bodies at surface of universe
$G = R^2/M * (c/s)$ where c/s is rate of spatial expansion at the universe surface

The three fundamental properties of a universe are mass M, radius R and acceleration of expansion at rate c/s where c is the fixed speed of light. c/s is the fundamental property of mass and speed of light where G is fixed for any universe of size R and M. Outward expansion c/s creates the equal and opposite force of gravity as -g, the rate of falling bodies at surface radius R.

That c/s represents the speed of light for any universe is easily demonstrated by letting R and M of the sun stand in for a small universe. Then $g=274$ m/s^2 per Newton's textbook $g=GM/R^2$, the instantaneous change in rate of falling bodies equal to the instantaneous expansion as redshift in the opposite direction $c/s=-GM/R^2$. Little g increases as GM/R^2 as R approaches zero, but c is invariant by direct measurement; therefore, objects on the surface of the universe accelerate towards the center the same as they do on all suns and planets. If c also increased as the little Einstein's claim Pound-Rebka proves, then the value of g would always equal -c/s, which isn't how we experience reality unless redshift if factored in. Therefore, rate of redshift is equal to the rate of metric expansion exerting the opposite force of gravity. This is not a theory or idea, but simply the meaning of long established and universally accepted observations. Ponder Figure 1 whenever your mind slips out of its orbit.

The change in g is the same as the rate of blueshift of light and of falling bodies within the universe as a whole between outer surface R and center R=0. The measured value of c must be invariant despite g being maximum at R=0 and minimum at R=surface (where outwardly measured big C is maximum due to accumulated redshift). g increases as distance to R approaches zero, but c decreases when adjusted for blueshift. The rate of change of blueshift is equal to rate of metric contraction; hence, c is constant using direct measurement and the definition of a meter.

CAUTION: Do not assume that the ratio of the sun's mass to 274 m/s is directly proportional to the invariant speed of light c such that mass M of the universe can be found by proportion. The density of the universe M/R^2 is many, many orders of magnitude smaller than the sun's density M/R^2, so the speed of light c that we measure is many orders of magnitude smaller than if the densities were the same. The theorem does not allow either universal M or R to be found because one must be known to find the other. It's an interesting thought that the sun might blow up with a big bang. Then what happens to G? If only the cosmologists knew what we know.

Geometric Axioms Of Foos's Theorem Of G

1) Property of mass is spatial expansion
2) Light is propagated away from mass by spatial expansion at rate c/s
3) Cause of gravity is equal and opposite reaction to force of expansion
4) Gravitational redshift = fractional rate of point to point expansion
$\left(\frac{\Delta v}{c}\right)$ or $\left(\frac{\Delta g}{c}\right)$ not Einstein's $\left(\frac{gh}{c^2}\right)$

5) Potential energy at surface of a universe is $E = mc^2$, where c/s is -g at the surface and the kinetic energy of falling bodies

The theorem and its implications should be self-evident, but official propaganda that photons lose velocity with mg has left the masses with serious brain damage. A back door approach should bring the problem to light, so ponder the following steps carefully. We show that Einstein's speed of light, constant or not, defines the universe and all objects within as black holes. They cite the 1960 Pound-Rebka as proof where c'=c(1+gh/c^2). Appendix B explains how relativity theory and entrainment theory claiming a variable c result in a universal black hole.

Given: Newton's $G = \frac{R^2}{M} g$ is equal to $G = \frac{R^2}{M}\left(\frac{c}{s}\right)$ at the surface of the universe where R is the universe radius. It cannot be argued otherwise because invariant c must be included in the definition of a universal constant and the degree of redshift at the surface corresponds to the change in velocity of falling objects. This is more easily understood by letting the sun be a small universe. Let the sun be a tiny universe of radius R and mass M; then plugging in known values of R and M for the sun we find g = 274 meters/s^2, the textbook value of g for the sun. If the universe by itself, 274 m/s is also the speed of light c equal to the opposite rate of expansion. So, c/s for the universe is the outward rate of expansion.

Since the invariant speed of light c represents c/s in Newton's formula when R is the surface radius of the universe, c/s equals the opposite value of g at the surface of the sun universe. The speed of light c is thus defined by the surface rate of expansion and redshift of the opposite g. The decrease in c where R'<R is owed only to spatial contraction as blueshift. Confirm from Newton's g that c/s = -g = GM/R^2, so the externally reckoned change in c equal to blueshift in the Pound-Rebka example can only be negative and not positive as wrongly calculated by the potential energy formula. This slowing of c with

blueshift is confirmed by the slowing of light as Mercury approaches the sun. The delay in light signals claimed by Einstein as owed to relativity is simply due to the contraction of space as blueshift which is equivalent to delayed light signals (clock speed). Any change in frequency or clock speed is reflected in the change in meter length by definition.

This confirms the fractional change in wavelength, light speed and clock speed over distance h in the Pound-Rebka results were correctly derived by Foos in Figures 1 and 2 in the Big Bang as $(1-gh/c^2)$, not the opposite $1+gh/c^2$ claimed by Einstein's worshippers. The expression using PE is incorrect, also, but that specific fakery wasn't closely examined then. The faulty PE term gh/c^2 (v^2/c^2) must be replaced by delta v/c to match the path of the falling crystal. Light expands as redshift with elevation at the rate opposite that of falling bodies and does not accelerate downward; the change in redshift being proportional to change in velocity, NOT change in potential energy PE. (Note all values of g in expressions over distance h are assumed to be the averaged g, g bar, over distance h. The faulty use of g in the Pound-Rebka article is obvious dishonesty meant to give Einstein false credit for the equations in Appendix A circled in yellow.

Important: What if your position is not on the surface of the universe where c/s ≡ g? Measured c is invariant, but g increases with diminishing R. Inferred c as c prime includes frequency change as spatial contraction; hence, c prime decreases at rate $c' = c\left(1 - \frac{\bar{g}h}{c^2}\right)$ as R approaches zero, if the fractional change in c matched the change in PE, which it could not. This is derived in Figure 2 in The Big Bang Boozle. The correct expression we now see should be $c' = c\left(1 - \frac{\Delta v}{c}\right)$ not $c' = c\left(1 - \frac{\bar{g}h}{c^2}\right)$. This will be narrowed down later, but the error is carried to illustrate the deliberately careless and shameless use of bogus mathematics to perpetrate the false notion that Einstein's relativity is again confirmed.

Net change in c incorrectly derived by Einstein, et al

In the sun universe, the initial c from Einstein's atomic cannons would be 274 meters/s as seen in Newton's G above where c/s is the rate of falling objects equivalent to g at the surface. The downward acceleration of imaginary photons per Einstein's proponents is

$$c' = c\left(1 + 2\frac{\bar{g}h}{c^2}\right) \text{ or } c' = c\left(1 + \frac{\bar{g}h}{c^2}\right)$$

depending on the interpretation of his theory explained in Appendix B. c is the initial muzzle velocity of emitted photons and c prime final velocity with declining elevation over distance h. The change in rate of velocity of emitted photons at R=surface of our universe sun wrongly derived from Pound-Rebka as $c' = c\left(1 + \frac{\bar{g}h}{c^2}\right)$ is 274 meters/s^2 (see Appendix B). Either depiction is an accelerating rate of downward fall equal to or twice the upward muzzle velocity of photons at 274 m/s on the universe surface; therefore, **the SUN CANNOT SHINE** at or beneath the surface of the sun or any size universe. Light at surface radius R or less cannot escape because Einstein's formula for attraction to mass as $c' = c\left(1 + \frac{\bar{g}h}{c^2}\right)$ or $c' = c\left(1 + 2\frac{\bar{g}h}{c^2}\right)$ are both equal to or greater than the rate that photons ejected at R or less will be pushed back into their cannons with the force of g and unable to escape. It gets worse since g increases as bodies fall towards the center. By mathematical proportion, light could not escape atoms at the surface or interior of any universe if photons fall at rate g as do golf balls (or twice that fast as many believe: see Appendix B).

Hawking's assumption on the speed of light c differs. He isn't foolish enough to contradict the invariant c proved by Michelson-Morley and verified many times over that the majority of Einstein and ether entrainment believers seem not to comprehend. Hawking is otherwise Einstein's devoted protege still regards light as photons shot upwards at c and pulled back like golf balls as the explanation for black holes. The Schwarzschild radius, another phony construction where the deflection of light is one radian which defines at what elevation downward velocity exceeds upward. This is a variation of Einstein's fake $2GM/R^2$ for deflection of light. His assumption is an invariant c where the variable rate of falling photons is equal to 274 meters/s at the surface of a sun universe and c is the invariant speed of photons emitted from Einstein's atomic cannons equal to 274 meters/s. An invariant c is slowed by gravity to produce a black hole. You said it! Hence, net muzzle velocity of light c would end up zero at the surface of the universe and less than zero as g increases with fall towards the center R=0. Einstein's universe must therefore meet the standard definition of a black hole on the surface of a universe and below. These models overlook the rate of expansion as c cancels the muzzle velocity of Einstein's cannons and that thus the SUN CANNOT SHINE. Since c is invariant and g increases as R decreases, light is forced back into atoms at a progressively greater rate the closer to the center of the sun universe or any region therein. Neither model works being based on air castles. This spatial expansion model based on the theorem of G is the only correct model for any universe of any size.

Foos's Theorem of G Lets The Sun Shine In

Let c accelerate upward as redshift from spatial expansion as correct inferred from the definition of a meter and an invariant c. This requires reversing the sign of Einstein's equation as correctly derived by Foos in Figures 1 and 2 (or using the proper notation h'-h). Then $c'=c(1-\bar{g}h/c^2)$, where the outward increase in c is equal to the downward increase in the rate of falling bodies or photons, and g bar is the weighted average of g and g prime. Carefully note that the locally measured value of c is unchanged because the fractional change in frequency is equal to the fractional change in c over h, thus any difference is offset by the definition of a meter. This accounts for the officially approved constancy of c established by Michelson-Morley and Louis Essen. Also, remember that

$c' = c\left(1 - \frac{\Delta v}{c}\right)$ is the correct expression, not $c' = c\left(1 - \frac{\bar{g}h}{c^2}\right)$

The equal and opposite force of spatial expansion in proportion to mass is responsible for the force of gravity and gravitational redshift of light at universal rate of c'/s and the equal and opposite value of g within the universe boundary. These facts were set out early in Figures 1 and 2 of the boozle book, derived from results of the Pound-Rebka experiment described in Appendix A before being overwritten to conceal evidence of Einstein's relativity fraud. The models of Einstein, et al, that depict light shot out of cannons at rate c either fixed or retarded by rate g result in photons stuck at their starting points. This theorem and all the claims herein are correctly derived from known, universally accepted laws of physics and observations that could be easily verified by an applied mathematician who had no fear for his job.

LET THE SUN SHINE IN

Simplified Universe Model And Revised Formulas

The earlier equations were based on Newton's and Einstein's expression for PE, potential energy despite change in frequency matching change in velocity and not the square of velocity. In the Pound-Rebka experiment the fractional change in frequency, (delta f)/f, is set equal to Newton's fractional change in gravitational potential gh/c^2 instead of g(bar)/c. As compelling as a change in PE might seem, the squares of velocity do not match the un squared velocity change of the falling crystal corresponding or the change in frequency, and are thus yet another deliberate and obvious deception. Einstein's dodge of meter definition is added to a pile of bogus claims that contradict Newton. The shabby omission of delta notation for h is also guaranteed to confuse his clueless apostles. Little g must be integrated over distance h and divided by h to get the corrected average. The overly complex relativity equation is presented as the solution to the difference in g over h when any competent scientist would recognize the solution is to integrate g over distance h or just use g1-go or $MG/R1^2=MG/Ro^2$. Pointing out this pervasive fraud in March, 2023, was the reason the wiki article displayed in Appendix A was overwritten.

It is much simpler and safer to replace R with the delta form where R1 is the lower elevation and Ro the higher. Subtract $MG/(Ro^2)$ from $MG/R1^2)$ as delta g. No integration skill is required now and the delta notation ensures the negative sign comes out correctly for change in clock speed r, length l and frequency f. Since potential difference is the basis of all energy change, the analogy of expanding ink in a bathtub of water aptly accounts for all the known laws of gravity where there is no resistance to unabated expansion of all points in mass. This is represented by the model in Figure 1. Rate of expansion c/s equals opposite rate of falling bodies, negative g, which is equal to GM/R^2. Using Figure 1 to establish your bearings, define a descending distance Ro to R1 instead of h, and see that (delta g)/g, is the fractional change in clock speed r, length l, and w wavelength, are all equal to $(GM/R1^2-GM/Ro^2)/GM/Ro^2$. They all represent the same equation because length is defined in terms of wavelength and/or clock speed, so change in one is reflected in change in all based on the definition of a meter. Unlike mgh, this is the correct expression for fractional change. The initial elevation must be subtracted from the final and divided by the initial. A professional follows standard conventions to prevent errors or confusion, contrary to Einstein's intentionally vague methods and excessively complicated technical language. Einstein defined delta f/fo correctly as change in f divided by inital f, then erroneously set this equal to delta gh/c^2 so that it would match the change in PE in the experiment, giving the false impression that delta f/fo and gh/c^2 are equal, but they are far from it.

In contrast, it is not confusing to stick with the correct model of spatial expansion derived from G as $R^2/M * (c/s)$ where R and M are radius and mass of any universe of size R and M and c/s the rate of outward expansion mandated by potential difference between mass and a vacuum. Unlike ink, matter is infinitely divisible in a vacuum and since there is nothing in an infinite void to resist expansion, the second law of thermodynamics is not violated; indeed, unabated expansion resolves the entropy paradox by restoring order to the universe that is thereafter continually lost in the realm of human experience. If you get lost, Figure 1 illustrates all aspects of the geometric dynamics of spatial expansion, so keep it handy to avoid confusion. The rate of expansion c/s is equal to negative g for falling bodies. The more meaningful and accurate version of fractional change delta c/s used in the wiki article is shown in the upper left. The change in light speed is not measured within the universal sphere except as gravitational redshift. I'd prefer using wavelength w than frequency f because the formulas are of the same form and sign, and change in wavelength translates directly to distance. Only this geometry accounts for the fixed measurement of c and meter length.

The maximum speed of light c for our expanding universe is the invariant and maximum rate of expansion at the surface. This is fixed everywhere within the universe since frequency change is offset by change in meter length. Surface expansion at c/s is the minimum rate of change in both outward expansion and the opposite rate of falling bodies. Note that wavelength of light and unit length increase at the same rate c/s such that c is a fixed constant determined by R and M. Little c added to accumulated redshift approaches maximum at the universe surface as the opposite -g rate of falling bodies approaches minimum. To avoid confusion, stick to subscript notation x1=xo where sub 1 is final and sub o initial.

So, delta c/s for our universe is easily seen as $GM/R1^2 - GM/Ro^2$ going away from the center as seen in Figure 1 and $-(GM/R1^2-GM/Ro^2)$ going toward center. The fractional change delta f/fo requires reversal of the sign because frequency increases toward the center. Delta c varies only as frequency change that affects the measurement of distance while the measured value of c is fixed. Wavelength w, length l, clock speed r and frequency now makes sense as they all boil down to the same parameter by the definition of a meter. Substituting the expression $(GM/R1^2-GM/Ro^2)/GM/Ro^2)$ instead of $(gh)/c^2$ leads to the (only) correct results instead of (gh/c^2) slipped in to wrongly tout confirmation of relativity. In particular, note that the erroneous increase in c with lower elevations claimed by the Einsteins (Appendix B) is $c'=c(1+gh/c^2)$ which when replaced by delta g readily simplifies to $c'=c+(delta\ g)$; however, since delta h is negative, the correct equation is $c'=c-(delta\ g)$ where c' is the slower speed of constant velocity c adjusted for blueshift with lower elevation. In other words, the value of c when adjusted for blueshift decreases opposite the increasing rate of falling bodies. The change in speed of light and value of g are simple opposites, a geometric fact, not a theory.

Keep in mind that any fractional change in velocity delta c/c of a falling object from Ro to R1, say 9.8m/s/c, is the same fractional contraction of space of 9.8 wavelengths/fo owed to blueshift. So, it is terribly easy to calculate the frequency shift over distance from any celestial body of mass in the universe. Never forget that delta c/s is positive outward rate of expansion equal to redshift seen internally and also equal to negative g, the rate of falling bodies. This is the combined rate of expansion between any two points in the universe and if close enough to be influenced by gravity the difference in expansion is the velocity of point to point gravitational potential.

Animated Model Of Sun Size Universe

A last shot at clarifying a few issues with the MSU math department. My last email probably left you a little confused, but this should clear it up. Please take seriously. This theorem is 100% correct and important to the extent a correct model of the universe should eventually overcome the Big Bang Boozle. The question still begging is why in the Pound-Rebka experiment was the fractional change in frequency equal to the fractional change in potential energy gh/c^2 (also equal to change in velocity squared over c squared). The article does not explicitly say that, but this is how it is usually taken; hence, the derivation of increased c by the Einsteins comes from the bogus fractional change = gh/c^2. How do fractional change in clock speed and frequency, delta f/f, equal fractional change in velocity squared ($PE/c^2=mgh/c^2=$ delta v^2/c^2)? These changes accompany only the change in velocity over distance h in the experiment, not change in velocity squared? As a mathematician, you should recognize an incorrect mathematical expression. The general relativity equations given (see Appendix A) are obviously fake. When do we learn that everything Einstein touches is a Rocky Mountain mess? Foos's Theorem of G illustrates the question in Figures 1 and 2. Why not let your own mathematical training and sense of reason be your guide? The standard definition of a meter uses a spectral line of cesium to measure the number of wavelengths in a meter. It also defines the unit of time between end points A and B. It is not elevation dependent and so proves there is no (local) frequency change or change in the velocity of light over Pound-Rebka's h where h=R1-Ro and R1 is the final, lower elevation. If there were any change, measured c would vary with frequency and there could be no fractional change in frequency or a standard meter. Our system of units would collapse.

A variety of experiments prove that to a fixed (remote) observer light is redshifted as it moves away from Earth (or any other massive body) and blueshifted going down. Neither velocity of light or wavelength change when measured by local observers attending a measuring apparatus, so the only (emphasis ONLY) possible explanation is that the apparatus itself has expanded. Space expands away from Earth propagating light waves with it. Gravitational redshift proves space itself expands away from Earth as bodies fall in the opposite direction. All gravitational redshift is a measure of the force of outward spatial expansion from a lower to higher potential, even the progressive intergalactic redshift. That is why the definition of a meter is the number of wavelengths or frequency of a spectral line or the time taken to go end to end. The definition of a meter also defines the length of a second. If the wavelength increases, the length of a meter increases even as its properties remain unchanged by expanding into a void. So if science and the definition of units has any fundamental, trustworthy meaning, every point in space expands away from bodies of mass as mass itself expands. Spatial expansion is the fundamental property of matter. This is the key to understanding gravity, length/time dilation and the true model of the universe, not Einstein's tinkering. Harken.

Pound-Rebka Formula For Fractional PE Fraudulent.

Now you see why the formulas for Foos's Theorem in Figure 1 have initial g in the denominator, not c squared. You have been duped by a Rocky Mountain lie. The correct expression for fractional change in wavelength or meter length is $(GM/R_1^2 - GM/R_o^2)/(GM/R_o^2)$. The sign must be reversed for frequency which increases as wavelength and length decrease. Figure 1 correctly states that the fractional change in frequency is equal to the fractional change in the rate of falling bodies. No squares of Einstein need apply. Put another way, the fractional change in frequency is equal to the fractional change in the length of a meter. Doesn't that make more sense being it's hard baked into the definition of a meter? Can you see it clearly now? If you can't, your're not a real mathematician or scientist. Since meters are defined either as number of wavelengths or the time taken going end to end, this is the true meaning of time and length dilation. They are really the same thing and were never properly explained by Einstein. Due to the contraction of space at a lower elevation, the signals received are delayed when observing the orbit of Mercury because half of a meter is half the time. Space and time signals are constricted on approach to the sun's mass, but the effect is only perceived by a fixed observer. It's important you understand this thoroughly.

The measure of c cannot vary because as frequency changes over a potential energy gradient, the length of a meter shrinks to the same degree so that the local length of a meter is unchanged. This is understandable in Figure 1 where the expansion of space results in the equal and opposite force of gravity in accordance with Newton's Third Law Of Motion. The force holding you to your chair is the difference in spatial expansion between different masses, but inside the box it appears you are falling. Here we easily understand that accelerating expansion of space means the maximum rate of expansion c/s at the surface defines the measurement of c inside the box because change in frequency is offset by change in length or distance, even as the acceleration at c/s into the void doubles every second and the external measure of big C geometrically increases. The unequal rates of expansion between different masses create the equal and opposite force of gravity and the outward propagation of waves of oscillating electrons at the frequency determined by their atomic shells. This is explained in detail in The Big Bang Boozle where the energy of universal expansion is $KE=mc^2$, $PE=gmc^2$, and the opposing nuclear binding energy $E=mc^2$. Since expansion is the equal and opposite rate of falling bodies, c/s approaches maximum rate at the surface while the opposite rate of falling bodies, g, approaches its minimum.

Yes, that's 100% correct. If internally measured c is fixed by rate of universal expansion at the universe surface, then negative c/s equals g, and that means all objects within must fall at rates greater than the speed of light relative to the universe surface; therefore, c is not the highest possible velocity but the lowest relative to the surface even though we cannot see the surface. Going away from the universal surface, if it were possible, the two values c/s and -g converge. Study the graph in Figure 2 and watch the animation at https://bit.ly/41gzHZi. The rate of surface expansion at radius R of any universe defines the speed of light c. Science is understanding. The speed of light cannot stand alone as its own creation, but depends on the size of a universe. Let a good mathematician understand this and earn his crown.

It is obvious that the velocity of a body falling past the surface of the universe exceeds the speed of light c and the two diverge progressively approaching R=0 as the fractional change in both rapidly diverge to maintain the measured constancy of c and length of a meter. The table in Figure 2 shows calculated values of c/s and change in length delta l above and below the surface at 10% of surface R intervals for our sun-size universe of 7.65*10^8 meters radius. This is shown by the brown column in the table and brown circle in the animation which is fixed as perceived inside the box yet expanding outward. The speed of light of the sun universe is 274 meters/second where maximum c rate of expansion and minimum g, rate of falling bodies, are equal just as g is 274 meters/second for the real sun, the outward expansion of space seen as redshift. The formula for G is c/s=G*M/R^2. Change in frequency, delta f/f, is the mirror image of bodies falling at an accelerating rate g, while the constancy of c is the yellow middle line between them; hence, the length of a meter and speed of light are fixed. If this offends your preconceived, Einsteinian dogma that c is the maximum universal velocity, your sacred duty as a scientist is to change your world view to match the facts of physical geometry. The change in frequency with elevation is offset by the change in meter length which fixes the measured velocity of c and the local length of a meter. The density of our own universe is very thin, the surface very porous, yet nothing exists beyond the surface except perhaps some form of energy like the CMB. Everything inside moves relative to others at less than c. They do not descend directly towards the center but with radial acceleration like we see in the arms of galaxies. Understand, these are fixed geometric facts, not anyone's theory or my idea. The solution is simple algebra only slightly more complicated than the Pythagorean Theorem if your mind can be freed from the mainstream's Big Bang science fiction.

I promised you an animation showing how these simple equations in Figure 1 define outward spatial expansion as redshift away from bodies of mass while the measurement of distance remains fixed inside the box. Figure 2 is a snapshot of the animation at https://bit.ly/41gzHZi and graph made from these formulas. They illustrate how accelerating spatial expansion causes the retrograde force of gravity and invariant measure of c. The little man perceives himself falling towards the sun's surface "inside the box" because the metric measure of expansion is fixed by meter definition, or you could also say that since everything expands outwardly with the box in proportion to mass, expansion is imperceptible inside the box except for redshift. The opposite is true if viewed outside the box where rulers expand with redshift and the man is actually being pushed outward at 274 meters/second/second just as if in an accelerating car. As complicated as these geometric axioms might seem, a good mathematician will easily recognize them as immutable geometric constants no less reliable than the area of a circle. Become as familiar with them as possible, and you will see the universe correctly for the first time in history. This is not a theory, personal idea, not relativity, but the fixed geometric relationship between spatial expansion, mass and gravity never before correctly explained. This is Foos's Theorem of G waiting for a few good mathematicians to recognize it for what it is. Please validate the theorem and let Montana State University take credit for training the mind that found the Holy Grail of cosmology. This is Foos's Theorem of G.

Unravelling The Pound-Rebka Einstein Hoax

This section is partly to correct the prevalent notion that c is variable over a gravitational potential derived from Einstein's phony fractional change in f, delta f/fo being equal to gh/c^2. (Laurence, Valev, heads up). The Google group may also consider this an elementary algebra lesson. For eight years university math and science, long days and top grades, I had these principles drilled into my head, so I've earned the right. If you're ready for the red pill, pay close attention. Every math and science teacher since the eighth grade taught me everything that follows. My ninth grade instructor, Mr. Bobinski would back me up even if he had to face down Einstein. You, too, can see through the Pound-Rebka hoax if you have ninth grade algebra firmly under your belt. You may need some supplemental images from the theorem; find them here: https://bit.ly/43AzNw8 :

From numerous exchanges with group members, some logged in Appendix B, it becomes apparent that most Einstein followers and even critics believe the Pound-Rebka experiment proved a variable speed of light according to the formula c'=c(1+gh/c^2) (or c'≡c(1+2gh/c^2). The equation derives from Einstein's fractional change in frequency, delta f/f erroneously set equal to gh/c^2. The only way the fractional change in frequency can be "proved" is by dishonestly setting delta f/f equal to gh/c^2, a deliberate Rocky Mountain violation of algebraic principles. The Big Bang Boozle delayed addressing this fraud directly until the last chapter. The many fallacies claimed to be inferred from the PR experiment are laid bare and the correct meaning of fractional change in frequency with gravitational potential derived within that context, so buckle up. Foos's Theorem Of G correctly states the unseen fractional change in frequency relative to the universe surface is delta f/fo = delta g/c, reduced from ((GM/R1^2)-(c/s)/(c)), the initial value of g being a negative c/s. However, this is correct only for objects in free fall from the universe surface at rate g=GM/R^2 which is perceived internally as c, the measured rate of universal expansion and fixed speed of light. This is a geometric principle for the universe as a whole but doesn't directly apply to the change in velocity between two internal elevations such as in the Pound-Rebka experiment. The fractional change in frequency is strictly the fractional change in velocity, f/fo=delta g/(GM/Ro^2) over distance h. If this correct change in frequency was correctly calculated, it would disprove Einstein's equations and fake theory, so it is made falsely equal to gravitational potential gh/c^2 to match Newton's correct formula for PE. This should be obvious to any freshman algebra student familiar with the rules, but that kind is not hired.

This section carefully uses elementary principles of algebra to show how the variable c as c'≡c(1+gh/c^2) was erroneously derived from Einstein's clever fakery, delta f/f≡ gh/c^2, which was falsified to match the Newtonian calculation for PE. The correct expression for fractional change in f is not gh/c^2 but delta g/go, a typical proportion of the kind introduced in elementary algebra. The original Pound-Rebka wiki article in Appendix A is used for reference because the same equations and relativistic claims are repeated in the bulk of literature on this subject, amounting to thousands of articles repeating the same false statements. They are deliberate misrepresentations key to the Einstein hoax, not the errors Prokaryotic claims to have fixed, yet if he had known what they were, he'd be writing this and not me. The original wiki article in Appendix A serves our purpose:

Starting with the second page of the wiki article, the topmost equation circled in yellow is presented as the relativistic formula for f (final as receiver) from f (initial, emitter). You, the mark, is expected to believe this oddly complicated formula corrects the Newtonian matter of g being different between the two elevations in the expression gh/c^2. Nobody could explain how this has anything to do with relativity or simultaneity. Any good mathematician will easily see that the difference in the two values of g is correctly overcome by the integrated average or the simple difference between the two in this case, delta g, equal to g'-g. Einstein's magnificent equation purporting to account for the difference in g is a big in a poke. It does nothing of the sort, but was crafted out of nothing for the purpose of creating this false impression. Rearranging terms, Einstein's fractional change in frequency delta f/f gives the result shown in red by Foos. Try it yourself and see that it is outside the range using g final and g initial shown in blue and can therefore be nothing but shameless fraud. The detailed derivation of this is found on pages 15-16 of the Boozle book. The most important matter revealed is the repeated tactic Einstein uses to promote his relativity hoax. At every turn, he simply tinkers with variables until they approximate a predetermined, incorrect result used to create the false impression of a more accurate answer than Newton's formulas would give. This isn't math or science. It's fraud. It's an unforgivable shame that the centers of influence would ignore such malpractice and further promote it as the foundation of physics.

Now skip to the formula v (approximately) equal to gh/c bounded by the yellow rectangle. This is another Rocky Mountain skunk. The velocity of a falling body is its rate of fall at R equal to g; that is, $g=GM/(R)^2$. It is not potential energy gh divided by c. The implication of this fake is yet again that Newton's formulas are somehow wrong and need fixing, but it's just the same Einstein phony trick over and over. Small gh/c is not Newton's formula for change in velocity over distance h, but is instead delta v = $GM/(R1)^2 - GM/(R)^2$ shown in blue and is NOT an approximation. We are not interested in v at any level when considering fractional change, but in delta v between R+h=R1 and Ro=ground level. To pass this lesson, you have to purge all things Einstein from your mind and replace them with honest mathematics. Follow me:

Change in frequency or velocity is not equivalent to change in energy. The pretense is needed to match the change in PE over h and falsely confirm Einstein's phoney fractional energy formula. You do follow me, right? Rearranging terms for delta $E/E \equiv gh/c^2$ gives $E'=E(1+gh/c^2)$. Don't blindly swallow, but being already knee deep in a pile of brain farts, let's pretend that fractional change in potential energy is also equal to fractional change in c and not delta c/co. This is where your formula for variable c comes from. You do see that now, right? I derived several similar expressions myself for change in frequency, length and c seen in Figures 1 and 2 of The Big Bang Boozle book and viewable also at https://bit.ly/43AzNw8 . I expressed my doubts then, but saved the discussion until now. But that's how I know how thousands of scientists and Einstein worshippers like you all came up with this fake expression for a variable c, $c'\equiv c(1+gh/c^2)$ contrived by Einstein to falsely switch the correct formula for frequency with another false formula for PE so the experiment would falsely confirm "relativity." Laurence wanted to use this formula to convince me that c is variable, but it's an Einstein fake as you should see now. Bobinski would throw you out of the ninth grade. Even though units cancel out, velocity on the left as c must correspond to velocity on the right, not $gh/c^2=v^2/c^2$. You all just flunked ninth grade algebra.

More important in the Boozle book is the fact, officially ignored, that based on the definition of a meter being independent of elevation, any change in gravitational frequency must by definition be reflected in meter length l and clock speed r. So you see where I've correctly set all three variables equal to this incorrect form such as $r'=r(1-gh/c^2)$ in Figures 1 and 2 of the Boozle book. In Figure 2, change in c is included because the measured value of c is fixed, but if redshift is taken into account, the speed is in fact variable but accelerates away from mass as **$c'=c(1+gh/c^2)$** if fractional redshift is added to the length of a meter. Delta R here is negative, so h is technically negative as well, so c' is going up, not down. This is proper algebra. Note in the book I questioned the validity of fractional change in frequency being equal to fractional change in PE, but wasn't ready to outright challenge it until it was time to derive Foos's Theorem Of G, where the complete laws of gravity follow directly from the cause that Newton and Einstein failed to recognize.

You get an A if able to follow this maze. Wading through the trail of Einstein's fraud detracts mightily from the truth, but then you need not look back. Neither fractional change in frequency or light speed is equivalent to fractional change in PE as any good student would know, c' is NOT equal to $c(1+gh/c^2)$. Mr. Bobinski would rap your knuckles. Unit analysis requires units must reduce to equal on both sides of an equation. Einstein cheats by using expressions where units cancel, but velocity over velocity on the left does not correspond to velocity squared ($gh/c^{\wedge}=delta\ v^2$) on the right, at least not in this case. The ratio of unit lengths cannot be equal to the ratio of squared units, nor can fractional change in f, delta f/fo be equal to $delta\ v^2/c^2$ as given in the article. Three apples out of four may be numerically equal to six oranges out of eight, but three out of four is not numerically equivalent to nine out of sixteen anything. Take everything belonging to Einstein and burn it. Review your ninth grade algebra and follow me while we correctly for the first time in history interpret the PR experiment using spatial expansion and redshift based on metric definition.

The Truth Is Easy Once You Know How

Using your own head, consider that the Pound-Rebka crystal accelerates during fall from Ro to R'. Final velocity v equals g'. We're not particularly interested in v, especially using Einstein's incorrect formula $v=gh/c$ instead of $g=GM/R^2$. We find delta v by proper means, $GM/R'^2-GM/Ro^2$. The PR experiment and others prove there is a gravitational frequency change proportional to g, so the fractional change in f, delta f/fo, is proportional to delta v/vo because frequency and motion are affected in the same way by gravitational force. This is not ordinary Doppler shift created by the relative velocity of a moving object, but the change in frequency equal in magnitude to the Doppler shift of a falling body. The correct formula for delta f/fo is NOT the algebraically incorrect and non equivalent gh/c^2 Einstein used to fake the Pound-Rebka results, but

Foos's Theorem Of G - Escape From A Black Hole

Delta f/fo = (GM/R'^2-GM/Ro^2)/(GM/Ro^2) = delta g/go

You don't have to be a genius to figure this out if you had Bobinski in the ninth grade. By its nature, gravitational redshift and rate of falling bodies are both proportional to massM and inversely proportional to R squared. A good math student knows instinctivelythat delta f/fo must be equal to g/go, and given the definition of a meter be confidentthatspace expands away from a celestial body at the same rate objects fall in theopposite direction. There would be no argument over that among honest professionals;hence, you know everything Einstein is a Big Bang Boozle. You've been Boozled,baby.Because spatial expansion cannot be directly measured, the existence ofgravitational redshift implies that bodies fixed in distance actually recede from each otherat a rate that produces the observed Doppler shift, but this recession is not directlyobservable because all points in space expand simultaneously in proportion to mass. Theexpansion is not perceived by material objects; therefore, it is obvious that light having no mass is propagated by the spatial expansion of matter. Redshift records the rate ofrecession from point to point as influenced by the density of matter. With uniformrecession from spatial expansion, distances "inside the box" are perceived as fixed.

Expansion of all points in the universe in proportion to mass means two objects appear fixed unless they are close enough for the acceleration of one to overtake the other as when an apple falls. Progressive blueshift matches accelerating contraction of meter length. The smaller object "falls" towards the larger as viewed "inside the box" because the accelerating expansion of objects near enough results in one overtaking another. The observed redshift is thus progressive like intergalactic redshift, but the cause in both cases is the difference in accelerating rates of expansion, one due to mass and the other total expansion of the universe over extended periods of time. There was no Big Bang and the raisins are not moving away from each other as humans would measure. The rate of universal expansion is not perceived point to point except over vast periods of time. Rates of recession are reflected in the Doppler shift that would be seen "outside the box" where all space is empty but light is propagated into it.

Can you see how you've been duped? There are serious problems with all these Einstein confirming experiments, and every one of his equations is both mathematically and logically unsound to an accomplished mathematician. Take Eddington's measured starlight deviation of 1.75 from gravitational deflection. Among other grave flaws in the starlight measurement and theory was the failure to take into account atmospheric refraction of the sun. Refraction shifts the Earth's image two minutes out of 24 hours. How many arc seconds is that? 1800 off the cuff? You tell me. So 1.75 arc seconds means the sun couldn't have a shred of atmosphere. That's why Louis Essen called Einstein a swindler. He was. There is no need to consult Pound-Rebka to verify the area of a circle nor to test the above equations comprising Foos's Theorem Of G. Use the correct formula Mr. Bobinski expects and see that the fractional change in frequency is the exact opposite of the fractional change in rate of falling bodies. You were once hopelessly confused by the culture of Einstein's science fiction. There is no hope for those who linger there. But now, armed with a sound knowledge of algebraic methods, you have a golden path to follow. Let me know how far you've come.

EUREKA!

Figure 1: Simplified Model Of The Universe As Correctly Derived By Foos. Simplified and straightforward formulas for g, frequency, wavelength and length contraction are also set forth. If you've had trouble getting it all straight, pondering this illustration should bring focus:

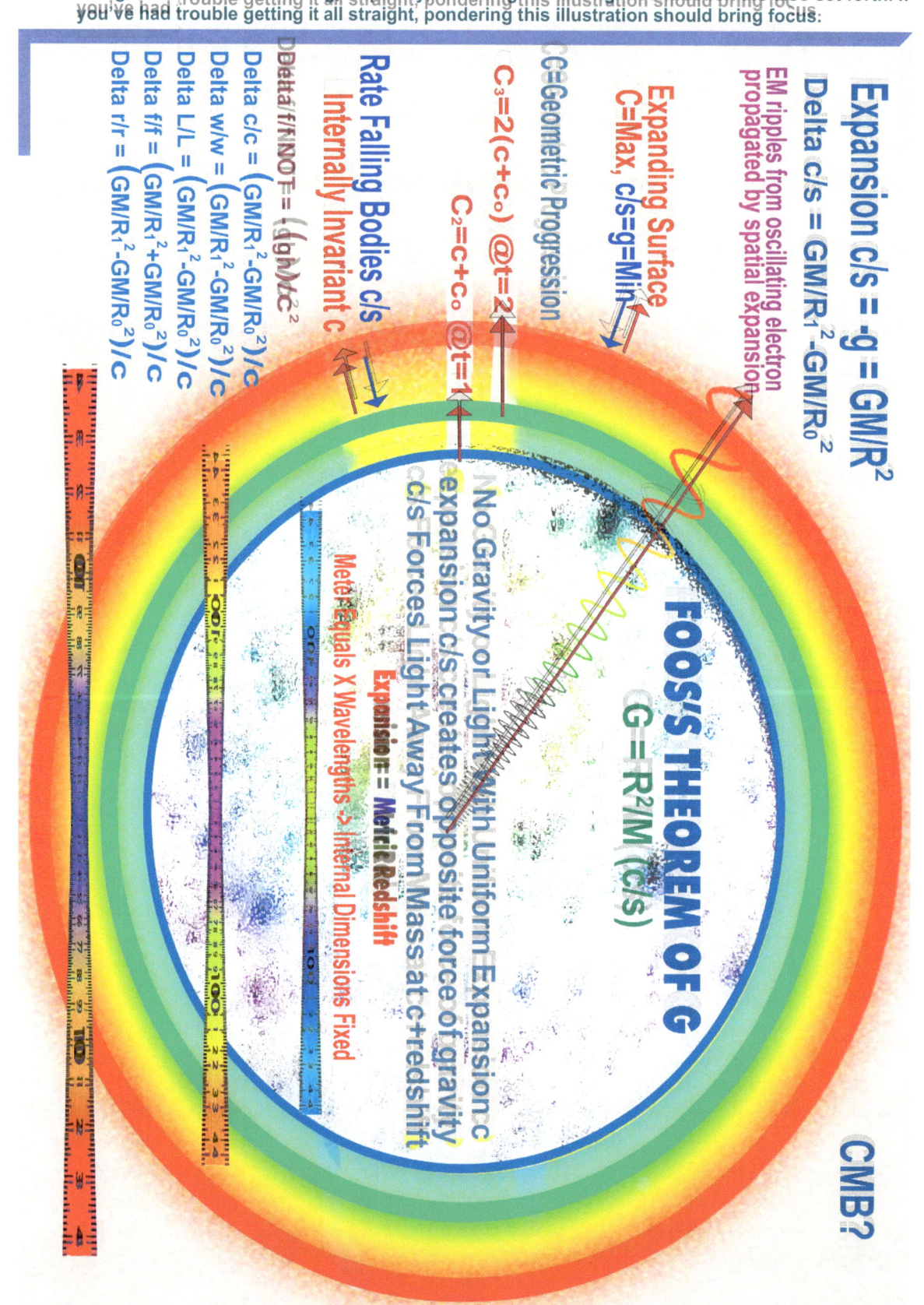

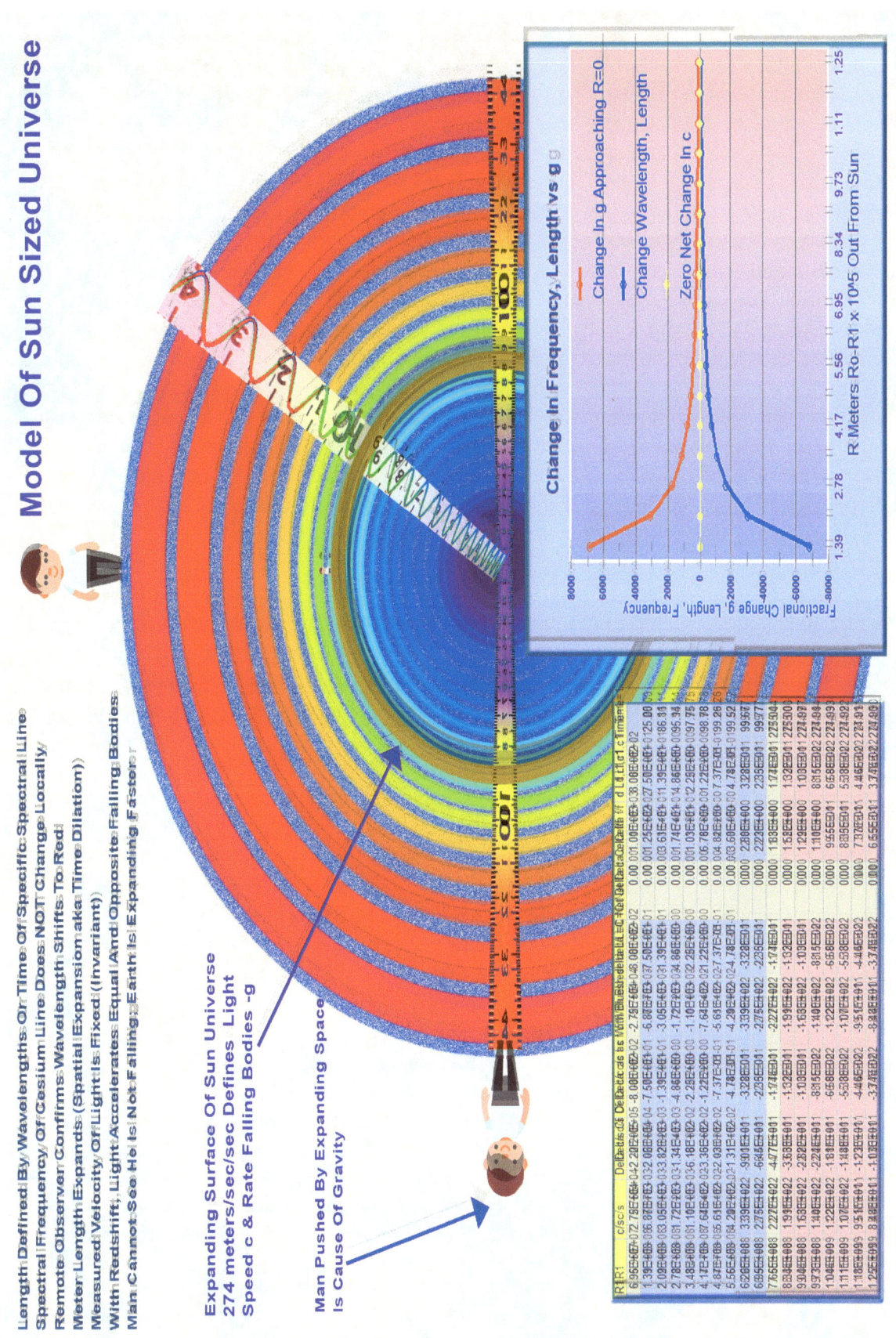

Figure 2: Animated Model Of Sun As Universe To Conceptualize Spatial Expansion And Gravity. Watch animated video via link https://bit.ly/41gzHZi.

Meaning Of Newton's G Proof of Einstein Fraud

This will explain in the most elementary and obvious terms why and how Einstein's relativity and the bulk of modern physics is a cult of scientific fraud. How is that possible when thousands of scientists scoff at that claim and perpetuate the substance and style of Einstein's fraud? Consider the success of cults like Scientology or certain extreme political movements. This example was prompted by conversation on Quora: https://bit.ly/46VONGX, see Figure 3. The discussion consists of the routine dismissal of Einstein critics as ignorant "cranks" usually motivated by anti Jewish bigotry or right wing fervor. There are some of those, but I know or care nothing about politics. I do have eight years of university math and science consisting of many thousands of the most difficult math and science problems that I was able to solve far more efficiently than my peers because of understanding honed by many years practice. A fraudulent equation I would recognize. In all explanations of Einstein like this, the language used is steeped in false intellectualism meant to make the reader feel less than capable of truly understanding the topic but too ashamed to say so. Variables are often left hanging without definition, omissions guaranteed to be marked incorrect in a classroom. If one dares point out these omissions or question deliberately obscured math, his comments will be deleted or his account closed. Despite the deeply pretentious language used by these "physicists," most of them have dismally weak academic credentials. This is the dishonest game played by modern physics and well explained in Bryan Wallace's book, The Farce of Physics. See https://bryangwallace.dreamhosters.com/.

So as I've done countless times, I commented on the Quora article to the effect that Einstein's equations and derivations violate the most basic laws of mathematics firmly established for hundreds of years and could not be considered anything but fraud. I pointed out specific examples backed by thorough familiarity with basic math principles first learned in ninth grade algebra and relied on to achieve perfect grades through two minors in chemistry and a master's in soils at Montana State. This comment and numerous others are quickly deleted by mods citing harassment or conspiracy theory. It isn't that anybody is harassed, it's that anyone who poses a genuine challenge is immediately banned. The only comments permitted are those who echo the praises of the Quora representative who dictates the conversation. This is how cults operate. Never in thousands of hours of university classes were my many questions not respected and answered. In my only upper division physics class, the instructor informed me that I was the only one with a decent grade. My criticisms are no less deserving.

It seemed odd that there was such a YouTube video, because I researched this topic and saw none except several of my own whose claims were not politically or racially motivated, but no less mathematically correct than any of the thousands of problems I solved at the university. Unfortunately, this isn't a topic of widespread interest, and most viewers are themselves Einstein groupies, so they didn't gain much traction. The average viewer is also not very proficient in basic algebra. So, I got this idea that I would do a video that uses a typical Einstein oriented Quora topic and using the most elementary algebraic principles demonstrate exactly how and why it is deliberate fraud. My comments on this one were also removed. In other words, the reason no proof of fraud against Einstein exists isn't because it isn't valid, but because the truth is improperly

Figure 3. Propaganda Praising Einstein's Theory But Ignoring Violations Of Basic Algebra, See https://bit.ly/46VONGX

11/10/23, 2:01 PM (12) If Einstein proved his theory of relativity, why then does science not universally accept this over Newton's version of gravity?

Quora    Search Quora   Try Quora+   Add question

If Einstein proved his theory of relativity, why then does science not universally accept this over Newton's version of gravity?

 Answer   Follow · 29   Request   ⓘ 💬 7 ⌄ ⋯

Viktor T. Toth · Follow   ✕
IT pro, part-time physicist · Upvoted by Frederic Rachford, PhD Physics, Case Western Reserve University (1975) and Jeremy L, PhD Physics, Singapore University of Technology and Design (2022) · 7y

Let me show you the equation of motion of a satellite around a planet under Newtonian gravity:

$$\ddot{\mathbf{r}} = -\frac{GM}{|\mathbf{r}|^3}\mathbf{r}.$$

Here, G is Newton's constant of gravity, M is the planet's mass, $\mathbf{r}$ is the satellite's position vector relative to the planet's center-of-mass, and the overdot represents differentiation with respect to time.

Now let me show you the same equation of motion with the lowest-order correction under general relativity:

$$\ddot{\mathbf{r}} = -\frac{GM}{|\mathbf{r}|^3}\left[1 + \frac{3v^2}{c^2}\right]\mathbf{r}.$$

That's it. That $3v^2/c^2$ term (v is the satellite's velocity, c is the speed of light), which amounts to a correction of about two parts in a billion for satellites in low Earth orbit.

Compared to this, the magnitude of the lowest-order correction due to the oblateness of the Earth is about one part in a thousand, which is a million times greater than relativistic corrections. So there really is no point to even worry about relativity theory before you learn how to expand the Newtonian potential in terms of spherical harmonics and then use, e.g., satellites to measure the spherical harmonic coefficients of the Earth. Once that's done, you can start worrying about the tiny, tiny relativistic corrections.

So it's not that science does not universally accept general relativity. It's that in most practical scenarios, the corrections due to relativity are so tiny, it is entirely sufficient to use the much simpler Newtonian theory (simpler in part because the mathematics involved is much more elementary) for practical calculations.

Remember, Einstein didn't invalidate Newton... his theory refined Newton's theory, and did so in a manner that Newton himself (who was very much troubled by the "action-at-a-distance" nature of his theory) would have appreciated.

Incidentally, you cannot prove a physical theory. Even if it is mathematically self-consistent, there's no guarantee that its predictions will agree with the real world. The predictions of Einstein's theory have been confirmed, sometimes with spectacular precision, but that's not the same as proof.

147K views · View 1,940 upvotes · View 19 shares

 Upvote · 1.9K ⬇   🔁 19   ⋯

Kip Ingram · Follow   ✕
PhD in Electrical Engineering, The University of Texas at Austin Cockrell School of Engineering (Graduated 1992) · Upvoted by Mark Jensen, Ph.D. Physics, North Carolina State University at Raleigh (1981) · May 16 · ✦

First of all, Einstein didn't - the experimental validations were done by others. Second of all, we don't "prove" theories of physics. We collect observations and see if they match the theory to within the error bounds of our observations. So far Einstein's theory has been validated in this

Related questions

Why do we still study Newton's theories even if Einstein has proven them wrong? Why don't w...

What is relation between Newton's and Einstein's theory of gravity?

How did the Einstein relativity theory destroy Newton's laws?

Did Einstein's General Theory of Relativity supersede Newton's Law of Gravity? If so, why...

Is natural selection a "theory" in the same sense as Newton's "theory" of gravity and Einstein's...

What is the difference between Einstein's and Newton's theory of gravity, and which one is...

How does Einstein's theory of relativity clash with Newton's laws?

Why was Einstein dissatisfied with the Newton's explanation of gravity? What caused him to go...

What is Einstein's theory of relativity explained simply (I am a 16 year old)?

How can we derive Newton's law of gravitation from Einstein's theory of relativity?

Add question

banned as harassment, conspiracy theory, or politics. So, in this YouTube video I will give this example of why Mr. Toth's explanation is outright, obvious science fraud typical of modern physics. Refer to this example, https://bit.ly/3tx5Y2S; and let's get on with it. Also, watch this YouTube video for illustrations https://bit.ly/47fNIsV.

First, Mr. Toth presents the equation of motion of a satellite by Newton. Why is this equation not Newton's and why is it deliberately misleading? Typically, equations are deliberately altered to make them impossible for an honest mind to understand; usually with many of the variables not defined. In this simple case the variables are defined but their true meaning garbled. On the left is r double dot, stated as the differentiation with respect to time of the satellite's position vector r. This obviously a deliberately pretentious and false perversion of Newton's formula for the acceleration of falling bodies, g, which is 9.8 m/s^2 on the surface of Earth. The independent variable is not differentiation and no such procedure could be shown in a real classroom. Of course, an orbiting satellite is not a falling body exactly, but in that case the acceleration is the same except it is orbital acceleration, not differentiation of the position vector with respect to time. It isn't harassment but correct to say that Mr. Toth like his peers is overflowing with a certain substance. NOTE that in the Pound-Rebka experiment v in v^2/c^2 is the instantaneous rate of falling bodies.

For more confirmation of that, we need only look at the right side of his equation. Why is there a negative sign in front? Unless bodies generally fall upwards, there is no reason for including a negative sign. This is fraud by small degrees. Then again, why is there an r cubed in the denominator. Mr. Toth could not possibly give a reason for that other than making the equation appear to represent something it doesn't in order to further bewilder an honest man. Nobody could explain the cube, not even Toth, but that's the point. Nor could he ever explain what is meant by a position vector. In Newton's equation, r is radius, distance from the center of the planet. There is no point in making r^3 absolute because just where and when could the distance from center be negative. Now you get it, right? Einstein and all modern physicists are simply frauds who misrepresent the facts as much as possible in order to confuse the minds of honest men. And again, what meaning does the solitary r in the numerator have? Can you tell me. Of course not, because you're being bamboozled. Those of you who passed elementary algebra can tell me that r in the numerator will cancel with one of the three factors of r in the denominator to give the only simple and correct Newtonian equation for the rate of acceleration of falling bodies, g=GM/r^2. Such pretentious wording and garbled equations are the core tactic of the physics cult of Einstein used to confuse the masses and justify their salaries.

But Mr. Toth is not yet done. He has yet to show you how Einstein's relativity provides a correction to Newton's formula even after twisting it beyond recognition. He does this by multiplying by the term (1+3v^2/c^2). He states that v is the satellite's velocity, but Einstein's equation actually states that g (r double dot) is the satellites instantaneous velocity at distance r, 9.8 meters per second where you're sitting now. This means that the numerical value of the independent variable g is equal to itself as the independent variable v. Any good ninth grade algebra student can tell you that such an equation violates basic algebraic principles and has no solution. The formula is fraud. Otherwise, Toth doesn't feel the need to explain the meaning of this term because it must be correct being Einstein; hence, it must be true. Notice that there are 1,900 up votes to this post even though not a single reader questioned this term or the deliberate bastardization of Newton's formula $g=GM/r^2$. If any did, they would have been removed from the conversion. Of course, we see the similar term $(1-v^2/c^2)$ in Einstein's general relativity where it refers to time and length "dilation." Toth not only has Einstein's sign wrong, but he's pulled the factor 3 out of a black hole. Nobody could explain what the 3 might stand for.

The equation is thus by the standards of math enough to prove fraud; but where does this $(1-v^2/c^2)$ come from? Explaining why Einstein's equations are fraud is a complicated exercise well covered in my book, but . But aside from violating fundamental algebra, it's extremely easy to show in this example why the use of such a term also violates elementary geometry thus on yet another count could not be anything other than fraud. Seeing this violation requires understanding what Newton's formula $g=GM/r^2$ actually means which isn't that difficult as Newton intended it.. This is a simple, fixed geometric relation like the area of a circle. If anyone told you that the area of a circle was not $pi*r^2$ you'd agree that the claim was fraud. Indeed, $g=GM/r^2$ is simply the formula for the diminishing interior density of a circle as it expands, i.e., the area of a series of concentric circles. I will show you why so you can seen clearly for yourself. Go ahead and draw a circle and fill it with a shade of gray. Let the density of gray represents the force proportional to density of a planet's mass. We know the force of gravity is proportional to mass, and so must be the rate of falling bodies. Hence, g is equal to M times a proportionality constant, say K; thus $g=KM$. Now increase the radius r such that the density of gray color is spread throughout the a larger area. Now the density of matter and thus force of gravity and g is diminished by the area of a circle. Hence, g must be equal to $KM/(pi*r^2)$, the area of a circle times a fixed constant of proportionality.

Not even Einstein could honestly claim that either K or pi could be variable, yet that is just what he and his devotees have done. Clearly K/pi is equal to G; hence, $g=GM/r^2$ which is nothing more or less than the density of an expanding circle. This is why the "experts," must constantly garble their equations. If they were properly understood then the fraud would be obvious to the ordinary bloke. Relativity is science fraud, and this example is sufficient proof of that. It isn't found in the literature because mainstream physics is run by a cult. The many other claims of the mainstream physics clique can be similarly exposed, but this simple example is enough to topple Einstein's air castle. I wanted this video to be as simple and short as possible to avoid risking confusion. If you have a good math background and want more detailed and comprehensive exposure of the Einstein fraud, you should read my book. It is well written, mathematically sound and heavily illustrated.

So, the main objective, an elementary proof of the Einstein's fraud and current state of physics theory is well accomplished. Anyone with a grasp of elementary algebra should see it clearly. I'm reluctant to proceed to the cause of gravity lest it invite confusion; however, if you're completely satisfied to this point, you may well wonder how Newton could so elegantly and concisely express the laws of gravity without knowing their cause. The cause is not difficult to deduce once Einstein and intellectual pretense is cleared from your head. Newton was handicapped by the undeveloped state of physical chemistry and statistical thermodynamics. Two main contradictions disprove the idea that gravity is a force between objects of mass. If it were, then the force would not be concentrated at the centers of mass as is well established. Once an object passed the surface, assuming a hole in the planet permitted, it's acceleration would slow on the way towards center as outer mass pulled on it. A different understanding of gravity is required.

Now consider a second fatal contradiction. Even Einstein couldn't miss the fact that the force of acceleration is equivalent to gravity. In other words, if you were in a rocket ship accelerating at 9.8 meters per second per second, you would weigh exactly the same as you do now on the surface of Earth where you perceive yourself sitting motionless. As counterintuitive as it is, you cannot be motionless but instead accelerating outward at 9.8 meters/s/s. Also consider that objects falling freely at the acceleration of g experience no internal force whatsoever so in fact they must be at rest. In both cases, it can only be that the surface f Earth expanding outward at the accelerating rate of g.

This you likely cannot accept because as you measure it, obviously the Earth is of fixed dimension and cannot be expanding. Here's how you resolve the matter. I all points in space expand in proportion to their mass, then the expansion of Earth would not be perceived because you also would be expanding in proportion to your own mass. Then the lack of force experienced by objects in free fall and the outward expansion of Earth makes perfect sense. The expansion of matter cannot be perceived or measured by objects of matter because everything expands at a rate proportional to its own mass. If you think about it carefully, you'll come see why that is true, why it must be true. When physicists say that space is expanding, deluded as they are, they don't mean space. They mean the distance between objects of mass. However, it is objects of mass that occupy space and so they themselves which must expand, not the space between them. Expansion of true space being matter itself, the cause of gravity and explanation of free fall make perfect sense.

Spatial expansion as the cause of gravity is also consistent with equilibrium thermodynamics as a special case involving no equilibrium. If you pour a bottle of ink into a full bathtub, the particles of ink expand outward with force (acceleration), at least initially. However, it is quickly diluted and slows as it encounters heavier concentration of water molecules until it reaches equilibrium. If, however, mass exists in a vacuum, which shouldn't be an argument, then there would be nothing to slow unabated expansion. And since it is mass itself which expands, all points within must expand such that expansion cannot be perceived except... expansion being the force of acceleration outward from the center of mass means that an equal and opposite force of gravity would be exerted on mass itself according to Newton's third law of motion. I conclude with confidence that if the fraud of Einstein were exposed, the cause of gravity being spatial expansion would become apparent to the masses.

Eureka!

One serious issue remains, that being the propagation of light. The true facts are easy to determine, but since they differ greatly from the Einstein mainstream version, special attention is required. In Einstein's famous 1919 prediction of starlight deflection, his theory was that light is deflected towards the sun by gravity, also a prediction of Newton's that Einstein claimed to have independently arrived at. This means light would accelerate towards mass like all objects, even though photons have no mass. This contradicts the findings of Michelson-Morley and others whose measurements of c show its speed is perfectly constant not only in all horizontal directions but vertically as well. Eddington's measurements inasmuch as they supported Einstein's predictions must have been faked. Indeed, light is bent about 1800 arc seconds by Earth's atmosphere. Since the sun's atmosphere is more dense than Earth's, Eddington's measurement of roughly 0.85 arc seconds assumes the sun has no atmosphere. Yes, there is a great deal of fanfare regarding the original and many repeated measurements of starlight deflection, but here we have already demonstrated clearly how Einstein's claims could not be other than fraud. Even if atmospheric refraction could be dismissed as the major cause of starlight deflection, experiments like those of Michelson-Morley and the very measurement of the speed of light entirely disprove the possibility of starlight deflection and acceleration of photons by gravity. If there was any change in the speed of light in a vertical direction, Michelson-Morley's highly sensitive interferometer would have shown strong diurnal patterns like the tides due to the motions of the moon and sun. This isn't to say there is no such deflection, only that it would have to be owed to atmospheric refraction. Finally, Louis Essen's measurements of c and later ones were not dependent on elevation. Light is not deflected by gravity or slowed on departure. Black holes are black for a different reason.

We don't actually need a theory or experiments to support it given the facts that gravity is owed to spatial expansion and the speed of light c is fixed in all directions regardless of the motion of source. Note the natural definition of light is a certain number of wavelengths of a certain spectrum of light. If the measuring apparatus is raised to higher elevations, the measured frequency of light and length of a meter is constant; however, due to spatial expansion, the higher meter must be longer relative to the lower by that amount. This we know must be true because gravitational redshift must coincide with spatial expansion. It may take some thought to incorporate this concept into your world view. This means that the fractional change in frequency, $(f_2-f_1)/f_1$, with distance from a large body must be precisely equal to the fractional change in g, the rate of falling objects, $(g_2-g_1)/g_1$. The false claim by Einstein et al that it is equal to initial f times $(1-v^2/c^2)$ (resulting from $f_2/f_1=gh/c^2$) involves the same violation of basic principles of math by Einstein as initially exposed. The proportion on the left must correspond another on the right. My books describe the fraud in detail, but it's much less pain if you take my word for it. It's important to clearly understand why accelerated expansion at rate c/s is perceived by matter as light driven outwards into space at a fixed rate of c. The fundamental property of matter is accelerated expansion proportional to mass, also known as dark energy. For the universe as a whole, $g=GM/R^2=-c/s$. Understand, if the Earth were the entire universe, the fixed speed of light c would be 9.8 meters/s as the acceleration of falling bodies at the surface would be 9.8 meters/s/s. This may be difficult to conceptualize at first, but it gets easier since it is a physical fact. The animations in some of my videos help understand it.

A final reminder about black holes. If light is not trapped by gravity, why are they black? This again is due to gravitational redshift, or better said blueshift. Larger bodies do produce a greater degree of redshift, but only because space is more constricted to begin with. Consider a source spectral line in red in deep space and a series of celestial targets of increasing mass. The line will be progressively compressed towards the blue with greater mass. When the object is a massive black hole, the frequency falls outside the visible range of light whether coming or going. The emitted frequency of black holes is in the cosmic ray range and fully invisible. That's all there is to it.

Definition Of Meter Proves Constancy of Light Speed c

My last video is highly recommended for understanding Newton's g=GM/r^2. Then it dawned on me that the masses are woefully ignorant about the definition of a meter, the fundamental of unit systems and which without Einstein illustrates the constancy of c to any rational mind. This then became a helpful exercise clarifying how spatial expansion explains true gravitational Doppler shift and the opposite rate of falling bodies. Perhaps one more desperate chop may save science after my death. Maybe a new angle exploiting Einstein's own example will brighten your understanding of gravity and the fraud of Einstein. This lesson could even benefit those hopeless boneheads like Lawrence who believe Einstein proved the speed of light variable in the Pound-Rebka experiment. This is eight years university math and science with near perfect marks talking; so listen carefully and fill your head with light. Watch this video: https://bit.ly/3PzydWR

The so called crisis in cosmology can be resolved using the simple definition of a meter. Real science arises out of the fundamental constants of nature, so pay attention: Originally, the meter was defined as the length of a certain bar of metal kept at constant elevation and temperature in a vault in Paris. This was a poor definition because the bar's length would vary under different conditions, and it would be impossible to make the original, undefied meter available to everyone. After Michelson and Morley proved experimentally the speed of light c was constant in all directions, the meter was finally properly defined as a given number of wavelengths of a certain spectral line or the time taken to travel that far; so definition of length simultaneously defines a unit of time. Since a sole vibrating electron in vacuum cannot vary, a meter would be precisely the same whether in Rome or on Mars or in motion. The meter is the foundation of all systems of physical measurements in science and the key to penetrating inconsistencies or fraudulent mathematics found in relativity. Sit back and learn.

The frequency of a spectral line and the length of a meter are blind to gravitational influence or motion, so c must be constant. However the measuring apparatus is placed, the color of the spectral line and the elapsed time will be exactly the same, so the velocity of c must be constant. No genius is required to see this clearly, so the credit given to Einstein for proof of that is fraudulent. Can you see that clearly?

The natural definition of a meter properly defines the fundamental meaning of time and space from molecules to cosmos. You cannot understand science without this as the foundation of your scientific mind. To penetrate the darkness further, let us borrow one of Einstein's famous but crude illustrations of a train passing by a stationary platform covered in the chapter on special relativity. This is what he used to "prove" his phony theory of relativity and the constancy of c as recorded in the history books. The deception stems from the failure to use Doppler frequency shift and the definition of a meter to correct for the velocity of the train, so let's examine the illustration carefully. You do now what Doppler shift is, right?

Carefully study the illustration of the uniformly moving train I will call inertial frame A. According to Einstein, as the headlight approaches the platform, light will travel the additional distance d by the train as reckoned by the stationary observer on the platform called inertial frame B. Einstein said inertial frame B measures the total distance traveled by light as c times time t, the velocity of the train times time t. Any fool sees clearly that the stationary observer must reckon the light travels the additional distance d in stationary frame B, but that would require c to be greater in B, which Michelson-Morley proved it cannot be! To resolve this dilemma Einstein devised a series of complicated algebraic steps to prove that the length of a meter and speed of a clock in moving frame A are less than in stationary B. The claim is that this proves c is constant and that simultaneity is entirely relative, not absolute. If you must believe this baloney because all the experts tell you it's true, then go away now; otherwise, recognize these claims as patently absurd. Relativity is the essence of scientific fraud. If the red pill is for you, follow this example to see the truth.

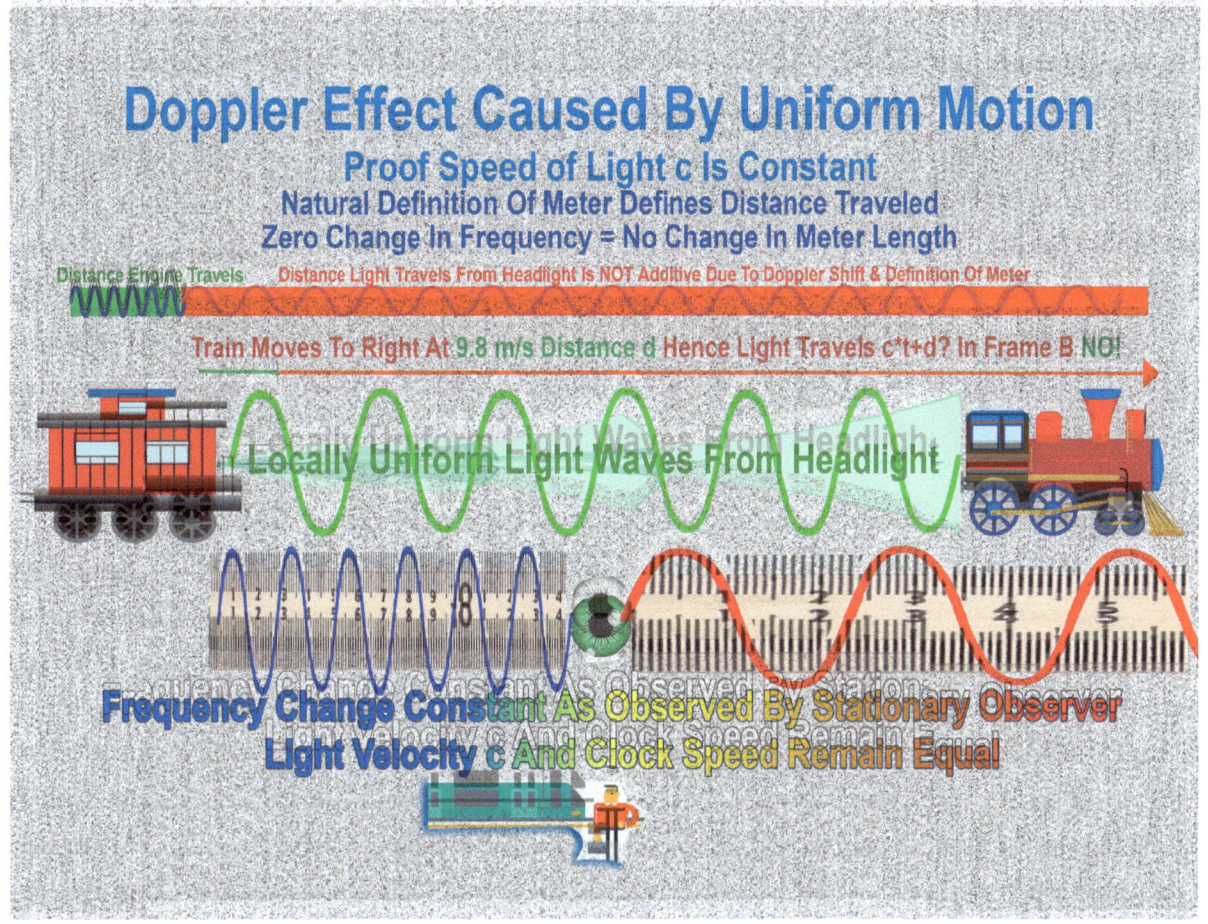

Figure 4. Moving Train Matching Eintein's Bogus Proof Of The Twin Paradox and Special Relativity.

Einstein did not consider Doppler shift or the definition of a meter in his theory, but we will use them in this example to make the truth self evident. Yes, it is true that in B's frame of reference, light from the train travels the distance of the red bar plus the distance the engine travels; however, experiment would prove there is no difference measured because the higher frequency of the Doppler shifted light results in a shortened meter that cancels out the distance d. The distance d is not additive for the speed of light given the definition of length and compression of light waves by Doppler shift. The train's motion causes the light waves to be compressed (blue shifted) in B's frame. B must use the meter shortened by Doppler shift to calculate the total distance traveled by light from the train which results in the same distance traveled as measured by A. The amount of blueshift is the exact distance traveled by the train in B. This isn't part of the distance covered by light as measured by either A or B using the meter corrected for blue shift. Do you comprehend clearly that this is why the velocity c is constant in all directions regardless of motion? The constancy of c is the natural result of the fundamental definition of a meter, not a result of Einstein's theories as mythology claims. You now understand it clearly without help from Einstein, right? If so, you may advance to accelerated motion. If not, review your freshman algebra and physics texts.

So one could argue that Einstein was right, that clock speed and length are slower (dilated) in A's frame relative to B, but that is not true in the cumulative sense sensationalized in the Twin Paradox. Note that there is zero point to point frequency change with uniform motion and therefore no ongoing change in clock speed or length. It is also arbitrary which frame is considered in motion, ironically Einstein's first postulate. The static, temporarily shortened meter is simply the result of initial accelerated motion by one frame or the other. The initial period of acceleration required to reach a uniform speed creates a transitory shift in clock speed and length in Frame A, but carefully note that when the train stops the compressed (bluer) wavelength is stretched the opposite direction and any effect on length or time cancels. The two twin brothers are the same age whenever they meet again. Einstein's twin paradox is an obvious lie believed by billions because the media sensationalized this among many false notions by Einstein. Reinforce your understanding that Doppler shift assures that the distance traveled by light from the headlight will be measured the same in Frame B or by any other body in any kind of motion. THIS IS WHY the value of c is fixed regardless of motion or direction, even in a black hole. If you do not accept this clearly as indisputable fact, you have stake in real science.

So, length and clock speed are NOT different between frames A and B as a result of uniform motion, the phony crux of Einstein's "special theory of relativity" that originally made him famous. Also understand why this absurd theory was called "relativity." It supposedly proved that the simultaneity of two events could not be absolute but must always be relative to the observer's motion or lack of it. You shouldn't need to be told this is untrue. If the light from two events meets midway regardless of motion, they are simultaneous; however, only an observer on the midline will see them at the same time. In the train example the claim was that the distance to midpoint will differ between observers A and B because of the additional distance the engine travels in Frame B. But again, there is no additional difference when Doppler shift is applied to the definition of a meter. Simultaneous events are absolute, no matter how infrequently they may be perceived as such, just as the sound of two cannons isn't heard at the same time for observers not equal distance from them. No perception of distance traveled by A within B's stationary reference translates to one clock ticking faster or slower as Einstein claims, nor does the meter vary in length during uniform motion except as a result of Doppler shift caused by acceleration.

Yes, only a change in velocity shrinks or expands a meter relative to a stationary point. The frequency of light reaching B from A's headlight is greater than at rest due to Doppler shift, but the "shorter" meter simply compensates for the speed of the train such that the distance and speed traveled by light is the same for all frames, an obvious fact that has nothing to do with Einstein's theory. The speed of light and distance traveled remains fixed thanks to the "relatively" shorter meter resulting from Doppler shift. The transitory relative compression of waves when the train initially accelerates is reversed when the train slows to a stop; therefore clocks and meters would all read the same when uniform motion ceases. Doppler shift is the principle behind the Highway Patrol's radar gun which is used to measure a uniform relative rate of motion. There is no variation in the velocity of light, definition of a meter or elapsed time or the results could not be valid. Change in length or clock speed is real, but requires a change in frequency. Ponder this until it becomes plain as day.

Wavelength is like a spring. If stretched by Doppler or gravitational effect, the time taken for light to reach end to end of a meter would stretch to match and the definition of a meter and constancy of c are both preserved. Locally, a meter would always be a meter. Doppler shift assures that a meter is always a meter and c is constant. Uniform motion does not affect clock speed or meter length except for the initial fractional change in velocity required to achieve uniform motion. Hearken! Einstein's first postulate was that it is arbitrary whether the train or the platform be considered in motion. Prominent scientists noted this would require every clock to run slower than all other clocks. Pointing out this obvious absurdity and the crooked algebra in his formulas, got them thoroughly blackballed as crackpots by a political cabal who has duped the public for over a hundred years. Here's your chance to escape the matrix. You do now see clearly that while perceptions will vary, simultaneity is absolute and neither c nor clock speed nor length are affected by uniform motion. The Doppler frequency shift simply corrects the distance light travels and makes it possible to calculate the train's velocity with a radar gun.

Michelson and Morley's experiments in the late 1800s proved long before Einstein what you now see clearly from the definition of a meter, that c must always be constant in vacuo. It should be obvious that if a certain spectral line is stretched or expanded by motion relative to an observer in another so called inertial frame, the time taken for light to travel the same number of wavelengths also expands; hence, the value of c must be constant. No genius is required to understand this. The natural definition of a meter assures that any changes in frequency due to uniform motion include simultaneous changes in both length and time, but no internal difference in clock speed or length takes place due to uniform motion. The Lorentz "transformations" claimed to represent such differences are false.

While special relativity deals with uniform motion, general relativity claims to generalize the absurdly false proof of time and length dilation from uniform to accelerated motion as if uniform motion were a special case of accelerated motion or if crackers were a special case of donuts. General relativity claims that the fractional change in frequency from gravitational acceleration corresponds to the ratio of potential energy to c^2, in other words, v^2/c^2 where v is the velocity of a falling body. Do you recall basic algebra and see this is algebraically invalid? It is pure nonsense to match fractional frequency change to v^2/c^2. Accelerated motion does cause length and time dilation, just not as Einstein's would have it. Now let's study the second illustration for an accelerating train. Note that as the train accelerates towards the platform the Doppler shift observed is not constant but progressively bluer from point to point. Even though the headlight does NOT change frequency relative to its local source A, stationary frame B clearly witnesses a progressive blueshift. Now there is indeed a fractional change in frequency exhibited by the changing motion of A which means corresponding, progressive change in meter length and clock speed relative to B. Study the illustration, not Einstein, and know how and why acceleration shortens the meter and slows the clock relative to B without affecting local clock speed or length. When you see this clearly, continue with this lesson.

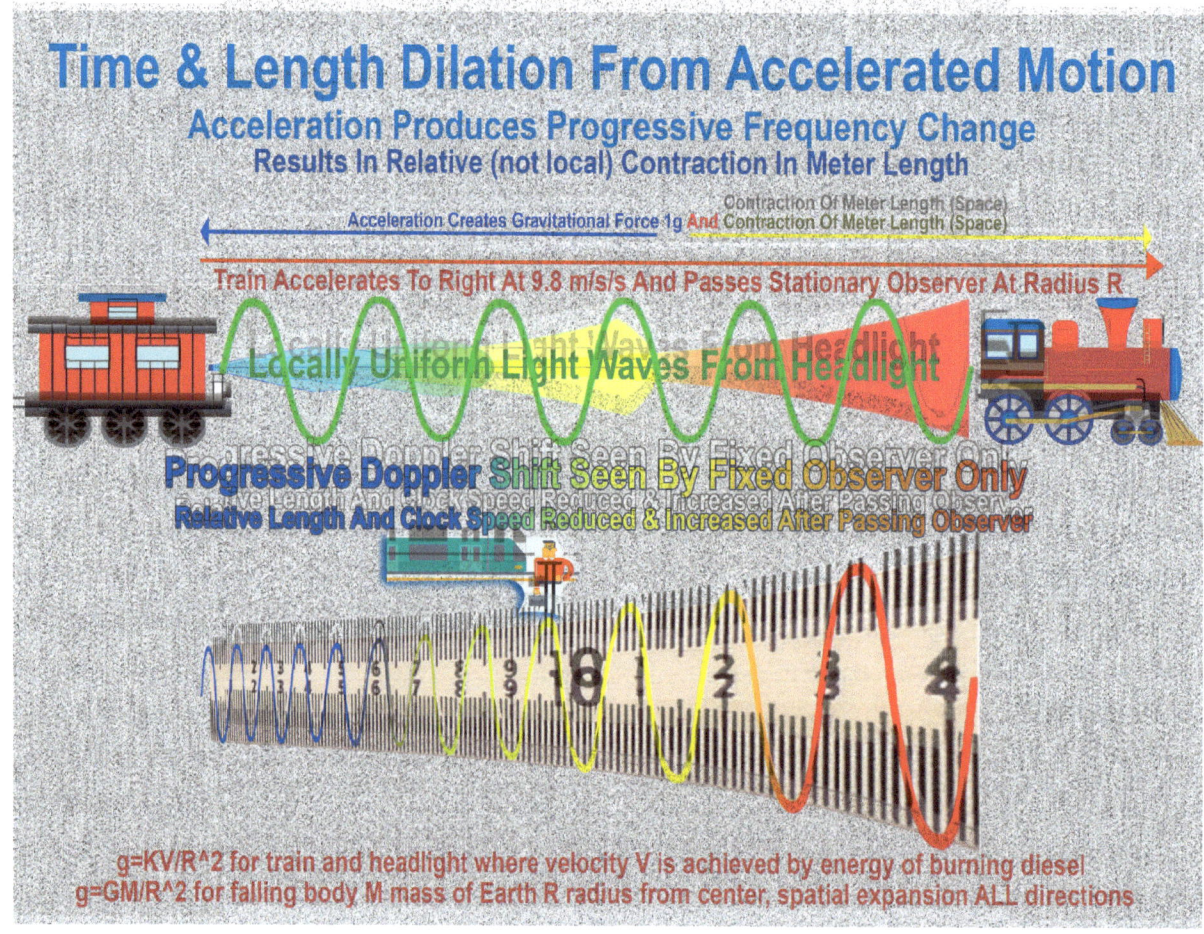

Figure 5. Accelerating Train Showing Progressive Blueshift And Length Compression

Now you can see clearly that acceleration produces cumulative point to point change in A's meter length and clock speed relative to B due to ongoing change in fractional frequency. The local definition of a meter requires that local measurements of time and space remain constant, but sustained acceleration towards an observer B progressively shrinks both time and space relative to the non accelerating frame. Obvious, isn't it, if you have a fundamental grasp of the meter. The velocity of c still remains constant. You do see this, right? Since there is zero change in material properties with changing meter length, the only possible conclusion is that all points in space can and do contract or expand with the sustained force of acceleration. You could kill the magic by pointing out that after passing the stationary observer in B that waves are progressively stretched the opposite way and the effect reversed such that once A becomes stationary with B the time lost in B is perfectly recovered. Given this symmetry, you might conclude the contraction of space and time is only academic. You'd be right about that except that acceleration due to gravity is in only one direction away from the center of mass and the effects irreversible. Einstein's empty postulate that acceleration and gravity are equivalent is only partially correct, isn't it? The train accelerates by means of the expenditure of fuel, but gravity is constant and requires no fuel. Gravitational length/time dilation are instead irreversible without the expenditure of more fuel than the universe could contain.

But most of all, the force of gravity is omni directional from the center of mass density. The question isn't what causes gravity, but what it could be that causes you to accelerate away from Earth's center at 9.8 meters/second. Spoiler alert! Objects in free fall experience zero force from acceleration. Like Frame B, they are stationary. It is YOU who experience the force of acceleration at 9.8 meters per second per second towards falling objects of any size. This is key to overcoming long embedded false perceptions. Nor are smaller objects attracted to large bodies or they would slow when passing the surface, but instead velocity increases as R approaches zero. It's not that the force of gravity acts as if it stems from the center of mass as Newton put it, but that the force of gravity does stem from the centers of mass outward. The only way to reconcile the contradictory perception is to accept the geometric axiom that the force of gravity is due to the accelerating expansion of all points in space away from centers of mass in proportion to mass density. This acceleration is not directly visible because all units of measure expand with it. It takes some time getting used to, but it is a fact that at its surface, Earth expands outwardly at the rate 9.8 meters per second due to its mass, while small objects in free fall are actually being overtaken by Earth's far greater rate of expansion. This as the cause of gravity is as much a fact of geometry as the Pythagorean Theorem. The progressive Doppler blueshift of acceleration equates to the opposite rate of expansion of space corresponding to the redshift of expanding meters. The Doppler shift of a falling object as measured by a radar gun cannot be due to an accelerating object as commonly perceived, because the surface of Earth like the train is what experiences acceleration, not falling bodies which are internally stationary. The addition of "nothing", as empty space means the properties of matter in expanding space are unchangeable, but the definition of a meter demonstrates there can be a relatively greater or lesser degree of nothing as all points in space expand simultaneously. Space expands outwardly from the center of Earth at the same rate objects fall in the opposite direction at the rate 9.8 meters per second per second on the surface of Earth. This is Newton's third law of motion: every action produces an equal and opposite reaction.

Now is the right time to ask again, how fast does Earth expand to make it appear that objects are falling freely when it is perfectly true they are stationary? This is again Newton's law, $g=GM/r^2$. It is immediately obvious that the degree of gravitational Doppler redshift away from Earth is equivalent to the opposite distance fallen, and that frequency shift must be independent of the presence of falling objects while must be zero gravitational Doppler shift relative to the co moving surface. IMPORTANT: True gravitational redshift is not perceived except as the Doppler redshift opposite the blueshift of falling bodies. That means frequency change is directly proportional to change in g with distance r. It may be difficult to perceive gravitational Doppler shift as a free standing wave because light must be bounced off a "falling" object to measure because this is how radar works, but careful inspection proves gravitational Doppler shift is propagated by Earth's expansion while independent of falling bodies . Of course, this is not the same as Einstein's calculation of gravitational redshift, a far smaller effect which would violate the real and well established laws of physics. Sorry, Albert.

Source Of Light Speed

Is it not perfectly clear now that spatial expansion is the fundamental property of matter and that spatial expansion propagates light? It is important to extend Newton's laws to their ultimate conclusion in the context of spatial expansion. Newton's $g=GM/r^2$ means that at the surface of the sun at distance R, that the rate of a falling body must be 274 meters per second squared, and if the sun were the entire universe, the speed of light must be fixed at 274 m/s/s. This relation holds true for any roughly spherical body of any internal density, from moons to galaxies to the universe as a whole. This means that surface g in the formula $g=GM/R^2$ must be equal to the opposite rate of expansion, c/s. Since the Doppler frequency change is equal to the opposite rate of falling bodies, it is strictly true that light is propagated by expansion and that therefore the accumulated maximum speed of light c is determined by the surface radius R of the universe as a whole. If the sun were the entire universe, then the speed of light c propagated away from mass at all points in space would be fixed at 274 meters per second. I see you squirming. How can that be? If light is propagated at 274 m/s/s and co moving with the sun's expanding surface, then the speed of light relative to the expanding surface would be zero, would it not? But unlike the sun, the universe has no continuous surface. The distribution of matter is extremely thin on the macroscopic scale such that all objects in space free fall opposite the propagation of light by surface expansion at rate c. This does not mean everything accelerates directly towards the center of the universe like a falling stone. Like celestial bodies in galaxies and the moon's orbit, everything accelerates at rate $g=GM/r^2$ but acceleration is normally orbital and rarely linear. That holds true even for subatomic particles where the vibrations of electrons are carried away from centers of mass at rate c. It may be that the mother of all black holes sits at the center of the universe, but most objects in the universe could not be visible to others since their relative velocities would naturally be greater than light speed. c is not the universal speed limit, but only another Einstein lie.

This concept is entirely correct but not easily grasped by the initiate. It is helpful to consider how things look "inside the box" where dimensions are fixed as opposed to how they would appear outside as all points in space expand. Inside the box at right, there is no perceptible change in dimensions by the reckoning of a standard meter. The surface of Earth appears fixed while stationary objects appear instead to be free falling. There is no contradiction when visualizing how things would appear "outside the box" at left. The expanding Earth overtakes the stationary "falling" man as light waves propagated by Earth's expansion are manifested as progressively increasing blueshift. Gravitational expansion is perceived as ordinary Doppler frequency shift. Expansion from point to point is canceled by the definition of space in terms of a meter. Of course, progressive intergalactic redshift is not due to the recessional velocity between galaxies which are on average fixed in distance, but due only to the difference in rates of spatial expansion between them. How surprising is it then that cosmologists infer from standard candles that the universe is expanding at the rate c/s without understanding that space itself is expanding and not the distance between galaxies? Nor is it any longer a mystery why the James Webb telescope keeps contradicting the big bang theory and showing it to be the fairy tale it is. Inside the box, the universe has always been the same size. Now you know where the constancy of c comes from as well as the fundamental property of matter being spatial expansion. Congratulations!

It should be mentioned that not only does the surface value of g define the value of c, but that all points in space expand with force at rate c/s; therefore, the waves from vibrating electrons are propagated away from matter at measured rate c. The energy of expansion

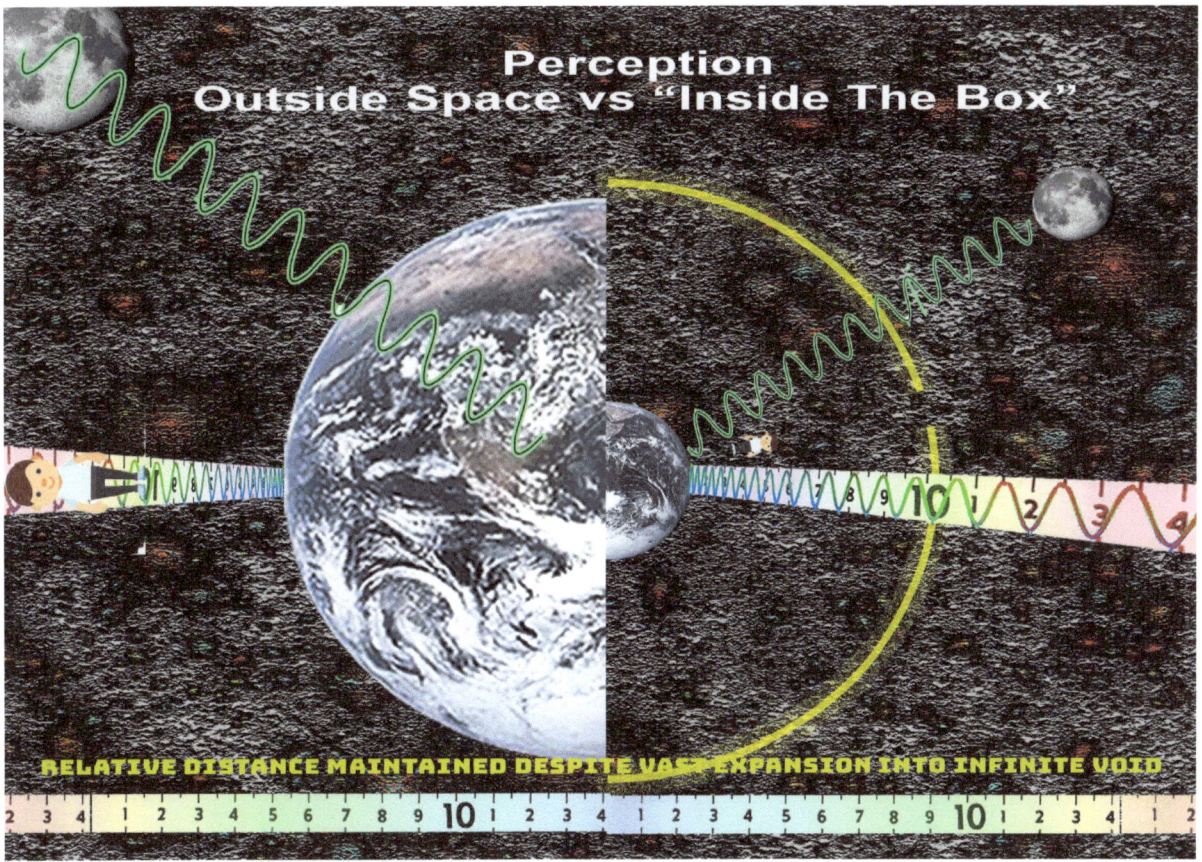

Figure 5. Percepton Of Falling Man Matches Expanding Earth If Inferred From "Outside The Box."

at the center of atoms also represents the binding energy of neutrons and protons equal Newton's energy of an inelastic collision E=mv^2 which on that level is the special case E≡mc^2. There is no conversion of mass to energy; indeed, fusion energy comes only from the release of an innermost neutron. Expansion is therefore the driving energy of the universe, the cause of gravity and the solution to the entropy paradox. The restoration of order lost through entropy according to the second law of thermodynamics is now fully explained scientifically for the first time in human history. This you now see clearly if your mind is entirely free of Einsteinian brainwash. Thank you.

APPENDIX A

Original Pound-Rebka wiki article deleted February 2023 after fraud revealed

Prokaryotic Capase Homolog, February, 2023, overwrites the fraudulent Einstein equation and other vital information exposed by Foos a week earlier: *"I have done a complete re-write of the Wikipedia article, which had a rather absurd number of inaccuracies."* Lone star Homolog has single handedly replaced the content and history of one of the most famous experiments in world history without consulting a single one of Earth's eight billion inhabitants, thereby erasing Einstein's general relativity and other evidence of scientific fraud.

The original is reproduced in this appendix to support references in The Big Bang Boozle and Foos's Theorem Of G.

Pound–Rebka experiment

The **Pound–Rebka experiment** was an experiment in which gamma rays were emitted from the top of a tower and measured by a receiver at the bottom of the tower. The purpose of the experiment was to test Albert Einstein's theory of general relativity by showing that photons gain energy when traveling toward a gravitational source (the Earth). It was proposed by Robert Pound and his graduate student Glen A. Rebka Jr. in 1959,[1] and was the last of the classical tests of general relativity to be verified (in the same year). It is a gravitational redshift experiment, which measures the change of frequency of light moving in a gravitational field. In this experiment, the frequency shift was a blueshift toward a higher frequency. Equivalently, the test demonstrated the general relativity prediction that clocks should run at different rates in different places of a gravitational field. It is considered to be the experiment that ushered in an era of precision tests of general relativity.

Jefferson laboratory at Harvard University. The experiment occurred in the left "tower". The attic was later extended in 2004.

Overview

Consider an electron bound to an atom in an excited state. As the electron undergoes a transition from the excited state to a lower energy state it will emit a photon with a frequency corresponding to the difference in energy between the excited state and the lower energy state. The reverse process will also occur: if the electron is in the lower energy state then it can undergo a transition to the excited state by absorbing a photon at the resonant frequency for this transition. In practice the photon frequency is not required to be at exactly the resonant frequency, but must be in a narrow range of frequencies centred on the resonant frequency: a photon with a frequency outside this region cannot excite the electron to a higher energy state.

Now consider two copies of this electron-atom system, one in the excited state (the emitter), the other in the lower energy state (the receiver). If the two systems are stationary relative to one another and the space between them is flat (i.e. we neglect gravitational fields) then the photon emitted by the emitter can be absorbed by the electron in the receiver. However, if the two systems are in a gravitational field then the photon may undergo gravitational redshift as it travels from the first system to the second, causing the photon frequency observed by the receiver to be different to the frequency observed by the emitter when it was originally emitted. Another possible source of redshift is the Doppler effect: if the two systems are not stationary relative to one another then the photon frequency will be modified by the relative speed between them.

In the Pound–Rebka experiment, the emitter was placed at the top of a tower with the receiver at the bottom. General relativity predicts that the gravitational field of the Earth will cause a photon emitted downwards (towards the Earth) to be blueshifted (i.e. its frequency will increase) according to the formula:

$$f_r = \sqrt{\frac{1-\frac{2GM}{(R+h)c^2}}{1-\frac{2GM}{Rc^2}}} f_e,$$

Pound–Rebka experiment - Wikipedia

Fractional change upper level using g @ Ro = 2.45298430548174E-15

Fractional change using integrated g bar = 2.45299295903122E-15 >Correct

Fractional change lower level using g @ R1 = 2.45300161255111E-15

Einstein's formula = 2.44249065417534E-15 OUT OF RANGE = BOGUS

where f_r and f_e are the frequencies of the receiver and emitter, h is the distance between the receiver and emitter, M is the Earth's mass, R is the radius of the Earth, G is Newton's constant and c is the speed of light. To counteract the effect of gravitational blueshift, the emitter was moved upwards (away from the receiver) causing the photon frequency to be redshifted, according to the Doppler shift formula:

$$f_r = \sqrt{\frac{1-v/c}{1+v/c}} f_e,$$

where v is the relative speed between the emitter and receiver. Pound and Rebka varied the relative speed v so that the Doppler redshift exactly cancelled the gravitational blueshift:

$$\sqrt{\frac{1-v/c}{1+v/c}} \cdot \frac{1-\frac{2GM}{(R+h)c^2}}{1-\frac{2GM}{Rc^2}} = 1.$$

In the case of the Pound–Rebka experiment $h \ll R$; the height of the tower is tiny compared to the radius of the earth, and the gravitational field can be approximated as constant. Therefore, the Newtonian equation can be used:

$$v \approx \frac{gh}{c} = 7.5 \times 10^{-7} \text{ m/s} \qquad \Delta v = \Delta g = \frac{GM}{R_1^2} - \frac{GM}{R_0^2}$$

Corrected by Foos

The energy associated with gravitational redshift over a distance of 22.5 meters is very small. The fractional change in energy is given by $\delta E/E$, is equal to $gh/c^2 = 2.5 \times 10^{-15}$. As such, short wavelength high energy photons are required to detect such minute differences. The 14 keV gamma rays emitted by iron-57 when it transitions to its base state proved to be sufficient for this experiment.

Normally, when an atom emits or absorbs a photon, it also moves (recoils) a little, which takes away some energy from the photon due to the principle of conservation of momentum.

The Doppler shift required to compensate for this recoil effect would be much larger (about 5 orders of magnitude) than the Doppler shift required to offset the gravitational redshift. But in 1958 Rudolf Mössbauer reported that all atoms in a solid lattice absorb the recoil energy when a single atom in the lattice emits a gamma ray. Therefore, the emitting atom will move very little (just as a cannon will not produce a large recoil when it is braced, e.g. with sandbags). This allowed Pound and Rebka to set up their experiment as a variation of Mössbauer spectroscopy.

The test was carried out at Harvard University's Jefferson laboratory. A solid sample containing iron (^{57}Fe) emitting gamma rays was placed in the center of a loudspeaker cone which was placed near the roof of the building. Another sample containing ^{57}Fe was placed in the basement. The distance between this source and absorber was 22.5 meters (73.8 ft). The gamma rays traveled through a Mylar bag filled with helium to minimize scattering of the gamma rays. A scintillation counter was placed below the receiving ^{57}Fe sample to detect the gamma rays that were not absorbed by the receiving sample. By vibrating the speaker cone the gamma ray source moved with varying speed, thus creating varying Doppler shifts. When the Doppler shift canceled out the gravitational blueshift, the receiving sample absorbed gamma rays and the number of gamma rays detected by the scintillation counter dropped accordingly. The variation in absorption could be correlated with the phase of the speaker vibration, hence with the speed of the emitting sample and, therefore, the Doppler shift. To compensate for possible systematic errors, Pound and Rebka varied the speaker frequency between 10 Hz and 50 Hz, interchanged the source and absorber-detector, and used different speakers (ferroelectric and moving coil magnetic transducer).[2] The reason for exchanging the positions of the source and the detector is doubling the effect. Pound subtracted two experimental results:

1. the frequency shift with the source at the top of the tower
2. the frequency shift with the source at the bottom of the tower

The frequency shift for the two cases has the same magnitude but opposing signs. When subtracting the results, Pound and Rebka obtained a result twice as big as for the one-way experiment.

The result confirmed that the predictions of general relativity were borne out at the 10% level.[3] This was later improved to better than the 1% level by Pound and Snider.[4][5]

Another test, Gravity Probe A, involving a space-borne hydrogen maser increased the accuracy of the measurement to about 10^{-4} (0.01%).[6]

References

1. Pound, R. V.; Rebka Jr. G. A. (November 1, 1959). "Gravitational Red-Shift in Nuclear Resonance" (https://doi.org/10.1103%2FPhysRevLett.3.439). *Physical Review Letters*. **3** (9): 439–441. Bibcode:1959PhRvL...3..439P (https://ui.adsabs.harvard.edu/abs/1959PhRvL...3..439P). doi:10.1103/PhysRevLett.3.439 (https://doi.org/10.1103%2FPhysRevLett.3.439).
2. Mester, John (2006). "Experimental Tests of General Relativity" (http://luth.obspm.fr/IHP06/lectures/mester-vinet/IHP-2GravRedshift.pdf) (PDF): 9–11. Retrieved 2007-04-13.
3. Pound, R. V.; Rebka Jr. G. A. (April 1, 1960). "Apparent weight of photons" (https://doi.org/10.1103%2FPhysRevLett.4.337). *Physical Review Letters*. **4** (7): 337–341. Bibcode:1960PhRvL...4..337P (https://ui.adsabs.harvard.edu/abs/1960PhRvL...4..337P). doi:10.1103/PhysRevLett.4.337 (https://doi.org/10.1103%2FPhysRevLett.4.337).
4. Pound, R. V.; Snider J. L. (November 2, 1964). "Effect of Gravity on Nuclear Resonance" (https://doi.org/10.1103%2FPhysRevLett.13.539). *Physical Review Letters*. **13** (18): 539–540. Bibcode:1964PhRvL..13..539P (https://ui.adsabs.harvard.edu/abs/1964PhRvL..13..539P). doi:10.1103/PhysRevLett.13.539 (https://doi.org/10.1103%2FPhysRevLett.13.539).
5. Hentschel, Klaus (1996-04-01). "Measurements of gravitational redshift between 1959 and 1971" (https://www.tandfonline.com/doi/abs/10.1080/00033799600200211). *Annals of Science*. **53** (3): 269–295. doi:10.1080/00033799600200211 (https://doi.org/10.1080%2F00033799600200211). Retrieved 2020-06-14.
6. Vessot, R. F. C.; M. W. Levine; E. M. Mattison; E. L. Blomberg; T. E. Hoffman; G. U. Nystrom; B. F. Farrel; R. Decher; P. B. Eby; C. R. Baugher; J. W. Watts; D. L. Teuber; F. D. Wills (December 29, 1980). "Test of Relativistic Gravitation with a Space-Borne Hydrogen Maser". *Physical Review Letters*. **45** (26): 2081–2084. Bibcode:1980PhRvL..45.2081V (https://ui.adsabs.harvard.edu/abs/1

APPENDIX B

This is a collection of excerpts of conversations with avid followers of Einstein and followers of the ether entrainment theory who make up 99% of those interested in or more or less proficient in math and physics. From this we get an idea of the contradictions between the special theory, examples of deliberate fraud or mistakes in his theories, and the completely unworkable consequences in the Standard Model. This helps provide a basis for proving Foos's Theorem Of G.

The Ether Entrainment Theory vs time dilation

SUMMARY OF ETHER ENTRAINMENT THEORY
https://bit.ly/3 mass, but the Planck energy hv of a wave is enough to knock an electron off its post without having mass. What binds modern Einsteins and Entrainers is the conviction that light has mass and is slowed by gravity at the same rate as falling bodies after being shot out of little atomic cannons.

KEY QUESTION: **HOW IS IT POSSIBLE FOR LIGHT TO BE SLOWED ON APPROACH TO THE SUN, BUT ACCELERATED ON APPROACH TO THE EARTH?** Astronomers observe a time delay as Mercury approaches the sun, the cause being cited as time dilation. That's proof of acceleration after passing perihelion. The Entrainers wear me out insisting it's all brainwash. Lots of reasons to believe it's fake. The starlight deflection boozle was definitely fake, the Hafele-Keating all too obvious, and the experiments too difficult to replicate by the ordinary Joe. But why then do they accept the Pound-Rebka experiments? Because the outcome fits their own preconceived notions. But too many astronomers all over the globe can check out Mercury's orbit to risk faking it. What time dilation doesn't do is explain the advancing perihelion because if the orbit was distorted by spatial contraction on approach, it would be expanded after passing. Time dilation would result in a perfectly symmetrical effect. It isn't possible time dilation is responsible for the advancing perihelion, use your head. But it is proof that light accelerates away from mass. Sorry, Einsteins. I'm going to hang with the spatial expansion model.

Another reason Einstein's formula is fake. He shows light slowing down on approach to the sun as $c'=c(1-v^2/c^2)^{0.5}$, the square root of the fractional reduction in velocity due to special relativity. The chapter on SR explains why this square root is fraudulent, and this example further confirms the charge. This isn't a uniform velocity issue, but a gravitational one falling out as a general relativity issue; hence, the correct expressions given by $(1-gh/c^2)$ or since g is not constant, more correctly $(1-v^2/c^2)$ where v is change in velocity over a given interval. Using the g bar average is also valid as derived in the Pound-Rebka chapter. The square root will not affect an experimental result and used only to fool idiots that special relativity (uniform motion) is valid. And, of course, either is in stark contradiction with both modern Einsteins and Entrainers who insist light is accelerated on approach to Earth. Any way you cut it, Einstein is the master liar.

So, the Einsteins insist light is slowed on approach to the sun and accelerated on approach to Earth. This schizoid brand of science is required to keep a government job as anyone who's had one knows. It's the only option for the many who can't do a real job. Their only method of countering debate is to make personal attacks if unable to outright ban critics. So, the Entrainers have good reason to mistrust the astronomers. The astronomers are lying to keep their jobs. Of course, if the astronomers are NOT lying, that means everyone else is, including Entrainers, with the sole exception of Foos. The only correct solution, assuming the astronomers are telling the truth, is the Foos equation in Figure 1 that reverses the + sign. Light approaching Earth, sun or any body of mass is slowed by $(1-gh/c^2)$, while outbound light is accelerated away at the opposite rate of falling bodies. Of course, ultimate accuracy is accounted for by calculating g bar as derived by Foos, not by Einstein's fake expression in the wiki article on Pound-Rebka before it was painted over by Prokaryotic.

But both camps also have to reject the invariant c proved by Louis Essen because all observers do not witness a variable c and frequency shift is a really poor substitute for change in velocity. So... both sides tacitly condemn the work of Essen, partly by denying the existence of a vacuum and partly by incorrectly citing experiments by Pound et al as proof light accelerates downward like all bodies of mass do. All direct measurements record a constant c. Even the Einstein's are forced to reject the original stated purpose of proving time dilation in the Pound-Rebka experiment. If they didn't, change in meter length would require light to accelerate upwards instead of Einstein's downward hypothesis. That's why time dilation was swept under the rug by Pound-Rebka. There were Einstein devotees, not Entrainers. Nor is Essen's invariant c ever mentioned, because frequency shift is by no means equivalent to velocity change. Sorry, boys, there is zero evidence that light follows the path of material objects.

No doubt the observed Doppler shift in Pound-Rebka is caused by the moving crystal and could be measured by any radar gun. But the point of the experiment was to show that the frequency of light itself changes by the same degree over a distance h in elevation. Because measures of c are invariant throughout, this means the Doppler redshift for light can only be due to spatial expansion. That works for a falling object, but velocity change for light whether due to frequency shift or not is indefensible. This harsh reality neither side can accept, so they must also reject clock speed and Essen's invariant c.

So Entrainers desperate to demonstrate a variable c resort to instruments like this, https://www.youtube.com/watch?v=7T0d7o8X2-E&t=222s. Eddington's estimate of starlight deflection seen from 93 million miles is valid, deflection by Earth's gravity would amount to not much more or less than a human hair over some miles (you do the math to be sure), not several inches over a few feet with such a flimsy gadget. In addition, Eddington and Einstein failed to account for the sun's atmospheric refraction which would have been a far greater source of deflection, they ignored the greater effect of refraction by the sun's atmosphere, the incorrect assumption that the light source was at infinite distance, and the shared elliptical orbit of all bodies of any mass. But here the Entrainers who despite Einstein commit the greatest fraud of all in attempting to demonstrate a variable speed of light. The only one you can believe is Foos.

Both camps fail with their explanation of the bending of light as owing to an invisible, massless ether entrained by gravitational force or mass. This massless ether is considered the source of an index of refraction that bends light, but mass is required to bend light, and bending light means it slows down opposite the theories of both that it is accelerated by mass. Note also that the frequencies of glass and gas both separate white light into a spectrum of color, but no such effect has ever been observed in a gravitational gradient because there is no refraction, nor would a stream of particles (or corpuscles) last very long if so restrained. But given an invariant c and the natural definition of a meter, the correct solution to the Pound-Rebka frequency shift matches the time dilation originally considered to be the point of it, no credit to Einstein. If clock speed and wavelength contract together, the Earth bound observation of Mercury's delayed approach to the sun would match all observations of interest, hence, light accelerates away from mass at (1-gh/c^2), not downward at (1+gh/c^2). This observation by astronomers is considered a trade lie by the entrainment camp. But time dilation can not be the cause of Mercury's advancing perihelion because the effect is reversed after perihelion and the resulting orbit perfectly symmetrical. Also, we earlier showed the expression for change in c as frequency shift is (1-v^2/c^2) Einstein's theories are a political hoax.

CONVERSATION WITH AN ETHER ENTRAINMENT THEORIST

The following exchange underscores the underlying error of both theories.

"No matter where they stand, the light going up is redshifted because it is slowed by gravity, and blueshifted going down. It has nothing to do with a vantage point."

Entirely not true. If the apparatus used to measure a meter is raised 22.5 meters, the frequency and length of a meter remain constant, a fact established by Louis Essen whose validated value of c is independent of gravitational potential. Light not slowed by gravity according to Louis Essen. Yes, the Michelson-Morley apparatus was horizontal, not to avoid a gravitational gradient, but to prevent the mechanical distortion you see in Grusenick's video https://www.youtube.com/watch?v=7T0d7o8X2-E&t=222s. The direct measurement of light is c invariant of any gravitational gradient it passes through. A variable c is not supported by measurement, nor is frequency change observed by moving a measuring apparatus through a gravitational potential. If space contracts according to the definition of a meter, only the Earth bound observer confirms both the frequency change and invariant constant c conforming to all observations and all experiments. Only spatial expansion fits the data..

"Light going up is redshifted because it is slowed by gravity." You have Doppler backwards. Review the accepted cause of Doppler shift in elementary physics texts. When objects accelerate away, light waves between them are progressively stretched (red shifted) by expanding space, just as a train whistle increases in pitch on approach. If motion slows as you claim, the result would be progressive blue shift. The only way to match the observation of redshift with Doppler effect is to reverse the sign, c' = c(1+gh/c^2) to c' = c(1-gh/c^2), then you have a match with Doppler shift and the correct mathematical derivations in Figures 1 and 2. Light accelerates upward, not downward. Anyway, redshift is not velocity, and an invariant c is measured from every point regardless of position. Those are the facts.

"A car is not shorter because it passes quickly." I'm pretty sure if you read the chapter on special relativity which in fact claims the car would be shorter, I say exactly the same thing. We aren't talking about cars in motion. That was Einstein's Twin Paradox hoax, not mine. We're talking about the behavior of light, not golf balls. The difference is crucial to understanding.

"Your definition in terms of wavelengths is wrong because it varies the unit of measure." Not at all. Quite the opposite, in fact. In order for a unit of measure to remain constant, it must conform to the definition of a meter, so if light waves are stretched by a gravitational gradient, the meter itself must stretch with it. Note that light frequency does not change when the apparatus is raised. If that were not the case, the natural definition of a meter would fail and the invariant speed of light established by Essen would be violated. The entire system of units would collapse. The spectral line of cesium used to define a meter remains constant with distance from Earth as measured by the apparatus. But the same fixed spectral lines are in fact redshifted by the reckoning of remote observers, the whole point of Pound-Rebka. If metric units are to remain constant, then by definition the upper meter must have expanded relative to the lower meter. Review your physics text and see that red shift means increase in velocity, not decrease. To be correct, you have to match all the observations.

"I don't know of any evidence for redshifts involving planets." The whole point of the Pound-Rebka experiment was to demonstration gravitational redshift and time dilation on planet Earth. Such experiments are not feasible on distant worlds, but why would there be a difference other than the value of g? Gravitational redshift of fixed stars is a fact of observation. Are you saying there is no gravitational redshift between here and the fixed stars? Cosmologists and astronomers say otherwise.

You asked me to look over Valev's change in light velocity with gravitational potential. Understand that when remarks become personal or accusatory then discussion is difficult. HOWEVER, in fairness to your offer, I scoped out articles by Valev to get a sense of his mindset. I settled on this one (see farther below). His primary focus is Einstein's bogus factor of two which wasn't the point of the original Pound-Rebka experiment, however interpreted. It must have been scrubbed from the wiki article decades ago like the bogus relativity equation Prokaryotic buried last month.

Please note in the attached PDF that numerous quotes by Valev show the downward increase of velocity of light at a variable $co=c1*(1+gh/c^2)$. The only exception (gone unnoticed) is by Fowler who alone correctly pegs both velocity and frequency as observed unchanged by the person in the elevator (local observer). Both frequency and velocity changes are ONLY perceived by the Earth bound observer as I've been trying to explain. The original point of the Pound-Rebka experiment was to demonstrate gravitational time dilation inferred by frequency change. Time dilation is not in dispute for physics in general, right? Consider the well known and accepted time dilation effect on Mercury's orbit. https://en.wikipedia.org/wiki/Shapiro_time_delay. This is a sound observation, not a theory or inference. Your firm interpretation of Pound-Rebka, sans clock speed, is only that light accelerates the same as golf balls towards Earth. I expect that extended discussion will only harden resolve, but there is a way to settle the argument for good.

Note that due to this time dilation effect, Mercury's motion is delayed on approach because the light signal is delayed. Delay means it is observed to be slowing down due to gravitational effect of the sun, then after passing it speeds up again. Do you see the problem? The velocity of light is slowed on approach to the sun, NOT accelerated. The object of science is to fit observations to mathematical principles.

So this is the billion dollar question. How is it possible that light approaching the sun's mass is slowed, but light approaching the Earth is accelerated? Of course, it isn't possible. The only solution is that the downward velocity of light in the Pound-Rebka experiment must also be reduced by $(1-gh/c^2)$, not increased by $(1+gh/c^2)$. Only be reversing the sign can the problem be resolved. That's why I had top grades in university mathematics. Can you answer the question?

Al

(To dodge the contradiction, he rejects the observation of Mercury's slowed orbit as an establishment lie because time dilation was Einstein's idea. Or was it? And he rejects the landmark invariant measure of light by Louis Essen, claiming light speed is instead variable and additive like all material objects, and that this variable ct was proved by Pound-Rebka.)

SUMMARY of Valev's article that follows:

Please note that numerous quotes by Valev show the downward increase of velocity of light at a variable $co=c1*(1+gh/c^2)$. The only exception (gone unnoticed) is by Fowler who alone correctly pegs both velocity and frequency as observed unchanged by the person in the elevator (local observer). Both frequency and velocity changes are ONLY perceived by the Earth bound observer as I've been trying to explain. Here is our source of conflict. Additionally, the original point to the Pound-Rebka experiment was to demonstrate gravitational time dilation inferred by frequency change. Time dilation is not in dispute for physics in general, right? So we're good so far? NOW, please consider the well known and accepted time dilation effect on Mercury's orbit.
https://en.wikipedia.org/wiki/Shapiro_time_delay

Note that due to this time dilation effect, Mercury's motion is delayed on approach because the light signal is delayed. Delay means it is observed to be slowing down due to gravitational effect of the sun, then after passing it speeds up again. Do you see the problem? The velocity of light is slowed on approach to the sun, NOT accelerated. If true, then the downward velocity of light in the Pound-Rebka experiment must also be reduced by $(1-gh/c^2)$, not increased by $(1+gh/c^2)$.

I expect that extended discussion will only harden resolve, but how can you explain that the stationary observer on Earth observes light accelerating downward when astronomer's confirm that light decelerates on approach to the sun in the case of Mercury? This is the crux of it, isn't it? I'll attach a PDF excerpt of Valev's essay. Fowler's comment is copied below. Maybe this helps bring us closer to agreement, maybe not, but it seemed like the answer can be settled this way. Either light accelerates away from mass as I deduce from clock speed and frequency, or the observation of Mercury's orbit is backwards.

Thanks,

Al

PAPER BY PENTCHO VALEV SUPPORTING ETHER ENTRAINMENT

Pentcho Valev explains how the fraudulent multiplication by two in Einstein's starlight deflection was exposed and disproved by Pound-Rebka. This error arises from the starlight deflection fraud explained in that chapter, but I was unaware that Einstein had used the same to predict the Pound-Rebka results which also proved it wrong. Einstein also predicted contracted clock speed (time dilation) inferred from the frequency measurement, but that was watered down later in the wiki article. After I pointed out the obvious fraudulent general relativity formula in February 2023 remaining in the wiki article on https://groups.google.com/g/sci.physics.relativity, the entire article was written over by a lone gunman for Einstein without bothering to consult any of the other eight billion inhabitants of Earth.

Mr. Valev's strident confirmation of Einstein's 2x fakery is appreciated, but the real purpose of pasting his web article below is to highlight the differing interpretations of how c varies in the Pound-Rebka example. The crux of the argument and unpleasant misunderstanding is revealed in the statement by Michael Fowler who alone provides the givens necessary for the correct mathematical solution to how c varies or not. This is the only interpretation that conforms to the unchallenged results of Michelson-Morley and Louis Essen as well as Newton's universal $G=R^2/M * (c/s)$ where all variables must be held constant. c is only a dependent variable inasmuch as it depends on mass M, but M is fixed. The constancy of G is also confirmed by assuming the sun or other celestial body is the entire universe and finding that c is consistently equal to g, the rate of falling bodies on the surface of any universe of any fixed mass and radius. The simultaneous interpretation of c/s as spatial expansion creating the opposite force of gravity, but that is the only correct algebraic solution to the Pound-Rebka results given an invariant c and the observations of time delay in Mercury's orbit. Only spatial expansion preserves the local system of units and establishes a cause of gravity.

The ether entrainment theory of my own was promoted between 2010 and 2013 but was playing the devil's advocate to gain a better understanding. Try on a few suits until you find the one that fits. Valev's discussion confirms the prevailing ether entrainment theory is identical to Einstein's adjusted general relativity where final c= initial c* $(1+gh/c^2)$. This mirrors Newton's flawed concept of light consisting of a stream of corpuscles with mass discredited over three hundred years ago by Huygens. The assumption that upward bound velocity of light is slowed by gravity the same as a ball thrown in the air is wrong. The time delay of Mercury's orbit proves the opposite effect according to $(1-gh/c^2)$, not $(1+gh/c^2)$. This is the crux of the underlying crisis in cosmology and the standard model as you can see. NOTE: To be strictly accurate, g must be replaced with g bar as described in the Pound-Rebka chapter, a trifle not accurately fixed by Einstein's relativity equation and since erased by Prokaryotic.

Ignored in both wiki articles is the initially stated purpose of the original experiment (take a look) to prove final clock speed is reduced with frequency as $r' = r (1-gh/c^2)$ as shown in Figures 1 and 2. It appears Einstein knew that this must be true if the properties of a meter were to be preserved. The corresponding contraction of length pops up separately as the stated principle of LIGO. Both of these were the original predictions of Pound-

Rebka but later weeded out. Frequency change with clock speed is impossible without spatial expansion or a meter no longer can be defined.

The present claims of both the Einstein and entrainment camps leave out the original length contraction predictions and portrays the Pound-Rebka experiment as particles falling downward or being equally retarded after being shot out of imaginary atomic cannons at rate c. The rate of fall and frequency change match the measured Doppler effect; however, a variable c is neither measured nor observed. The inference that frequency change is equivalent to velocity is valid for golf balls and protons, but not for light. This is wrong. It is contrary to all experience that light from galaxies billions of light years old would still be measured at a fixed rate c. There is no possibility that an ether exist with a refractive index that slows the speed of light it would be measurable with Essen's apparatus. NOTE that in special relativity, there was no ether and c was invariant, but in general relativity the ether is reinvented to slow and refract light the same as glass. This is forbidden by the measurements of Essen and Michelson-Morley showing an invariant c unaffected by gravitational potential.

The great unwashed are still convinced Einstein proved the constancy of light speed independent of source velocity and that he disproved the existence of an ether. YouTube is crammed with such videos as this:

https://www.youtube.com/watch?v=V7vpw4AH8QQ, but he actually reversed those claims in his final theory. So, which Einstein can we believe? Use your head. If an ether were to bend light, it would have the viscous properties of a fluid and retard the approach of light and the rate of falling objects opposite our experience. A golf ball's rate of fall would be slowed instead of accelerated. It's entirely absurd. Einstein's postulated ether in GR has no mass so as to sidestep that contradiction, but then if it had no substance it could not have an index of refraction. WRONG, Albert. Again. The redshift of upward bound light does not account for the measured invariance of c. Spatial expansion does. This is why the original objective of Pound-Rebka was to match increased frequency with reduced clock speed, and why length constriction pops up in the LIGO experiment. These two effects were necessary to explain the results of Pound-Rebka, but got lost in the shuffle. Einstein's notion of clock speed was also the point of the Hafele-Keating experiment touted as a marvelous confirmation of relativity.

Redshift is an accurate account for the changing motion of material objects, but not the path of light of invariant speed. Any high school physics text plainly demonstrates the opposite. If points in space accelerate away from other, they are redshifted and conversely progressively blueshifted when moving closer. If light did indeed have mass, its acceleration would be measurable, but c is invariant, and when observed indirectly it accelerates away from fixed stars, not toward them exactly as the orbit of Mercury demonstrates. These contradictions are not overcome by casually attributing them to redshift or time dilation as done.

The ether entrainment theory is an embarrassment. Grusenick's video made to illustrate the effect was the last straw: https://www.youtube.com/watch?v=7T0d7o8X2-E&t=304s. Several of these were built by a team of scientists promoting an ether model almost identical to mine. Going by Eddington's measurement of starlight deflection that made Einstein famous, the deflection by Earth wouldn't differ appreciably by from a human hair seen over the distance of miles, not several inches over every 45 degree turn of a four foot chunk of plywood. Gravitational torque on the mirrors is painfully obvious. Yes, Michelson-Morley's apparatus was horizontal and floated in a vat of mercury, not to avoid a gravitational gradient, but to eliminate the gross mechanical distortion demonstrated by Grusenick's contraption. There is no ether, only empty space.

History seems to have long forgotten the 1600s drama when Newton's corpuscular theory was rejected in favor of Huygen's wave model. Huygen's was right, light has no particles, no photon, and there is no ether or refractive index to guide their path. The invariant speed of light demonstrated by Michelson-Morley would have held true even if it were feasible to rotate the instrument without distorting the mirrors. How can you be sure? Because an accurate, invariant speed of light independent of gravitational potential was first demonstrated by Louis Essen in 1946. Nobody has ever contradicted his results, https://bit.ly/3YVQFee. If sanity is compatible with science, Mr. Essen's invariant c has the final say. It's important to note that he wrote two detailed essays pointing out the absurdities in Einstein's phony science. He also vigorously objected to the atomic clock he himself invented being used to validate Einstein's quack theories.

The experience with the Google physics group was an unpleasant but learning experience. In general, vigorous objections will be made to Essen's definition of constant light speed because "there is no such thing as a vacuum." Einstein's general relativity requires a massless ether with a refractive index to slow and the path of light, but particles of mass are accelerated by gravity, not slowed. The group majority consists of Einstein clones incapable of serious debate but only personal attacks. If Google were not hosting the group anyone criticizing Einstein would be banned instantly for promoting conspiracy theories. You soon sense that nobody in the group truly buys Einstein's theories, but defends them in the interest of keeping their government jobs.

So, then, if we assume the invariant c determined with accuracy by Louis Essen and accepted by all physicists of sound mind, then apply it to the Pound-Rebka experiment, we find that the demonstrated changes in light frequency in a gravitational gradient prove by definition that meters expand not on a local level but by the observation of fixed observers. The only way that c varies is by spatial expansion with redshift, but c remains invariant by both measurement and definition. It's remarkable to me that this unavoidable conclusion has never been grasped, but we don't live in a sane world. You object because nobody else supports my model? If they did, why write The Big Bang Boozle?

Hopefully, the detailed explanation in the book gives you enough context to understand Valev's following essay. The contradictions disappear nicely by reversing the sign in $c' \equiv c(1+gh/c^2)$ to $c' = c(1-gh/c^2)$ where c' is the final velocity of a downward light beam with final frequency $f = f(1+gh/c^2)$. If c is locally invariant, it's apparent that light and space accelerate opposite the rate of falling bodies. With no imaginary ether to explain and by application of the proven definition of meter length, we see clearly that spatial expansion away from mass matches the rate of falling bodies and that all physical laws and observations finally conform to an expanding universe in accord with Newton's third law of motion. There's no need to expound on this because the book says it all in many different ways.

In the following article by Valev, most references fail to correctly peg the invariance of both frequency and light speed by a local observer. Examine the one by Fowler framed between a series of * symbols to get your attention. Fowler alone correctly frames the problem in terms of locally invariant c and frequency both in contrast to the Earth bound observer who alone measures a frequency shift. If you set the problem up correctly with these givens as in Figures 1 and 2, you arrive at the fact that c only varies by remote observers as $(1-gh/c^2)$, the opposite of Einstein's $(1+gh/c^2)$. That's why I earned the top grades in university math. That's why observations of Mercury's orbit demonstrate time delay on approach to the sun. Light slows on approach according to $(1-gh/c^2)$, opposite the gravitational acceleration of the planet. Likewise, clock speed on Earth is retarded due to greater mass than on Mercury in accord with Figure 1; therefore, the observed speed also slows on approach to Earth at rate $(1-gh/c^2)$, not $(1+gh/c^2)$.

EINSTEIN'S RELATIVITY INCOMPATIBLE WITH GRAVITATIONAL REDSHIFT

Pentcho Valev (undated, https://bit.ly/3yX1jXf)

https://fr.sci.physique.narkive.com/nniOZxfL/einstein-s-relativity-incompatible-with-gravitational-redshift The Pound-Rebka experiment has actually confirmed the prediction of Newton's emission theory that, in a gravitational field, the speed of light varies like the speed of ordinary falling objects (same acceleration), and refuted the prediction of Einstein's relativity that it varies twice as fast as the speed of ordinary falling objects (if, for light, the emission theory predicts acceleration g, general relativity predicts acceleration 2g).

FOOS: No, the experiment confirms that light accelerates opposite g. Read on...

http://courses.physics.illinois.edu/phys419/sp2013/Lectures/l13.pdf

University of Illinois at Urbana-Champaign: "Consider a falling object. ITS SPEED INCREASES AS IT IS FALLING. Hence, if we were to associate a frequency with that object the frequency should increase accordingly as it falls to earth. Because of the equivalence between gravitational and inertial mass, WE SHOULD OBSERVE THE SAME EFFECT FOR LIGHT. So lets shine a light beam from the top of a very tall building. If we can measure the frequency shift as the light beam descends the building, we should be able to discern how gravity affects a falling light beam. This was done by Pound and Rebka in 1960. They shone a light from the top of the Jefferson tower at Harvard and measured the frequency shift. The frequency shift was tiny but in agreement with the theoretical prediction. Consider a light beam that is traveling away from a gravitational field. Its frequency should shift to lower values. This is known as the gravitational red shift of light."

FOOS: No!. The rate of a falling object is measured by progressive Doppler blueshift, but light itself having no mass is undergoes no velocity change, nor does the frequency shift itself account for a change in velocity, but is always c. The upward, outgoing redshift of light is Doppler acceleration only in the sense of spatial expansion since c is invariant.

http://www.oapt.ca/newsletter/2004-02%20Newsletter%20Searchable.pdf

Richard Epp: "One may imagine the photon losing energy as it climbs against the Earth's gravitational field much like a rock thrown upward loses kinetic energy as it slows down, the main difference being that the photon does not slow down; it always moves at the speed of light."

FOOS: That's is CORRECT. It is a fact that c is invariant when measured. It cannot be variable unless in terms of metric definition, and then it accelerates away from Earth.

http://www.amazon.com/Brief-History-Time-Stephen-Hawking/dp/0553380168

Stephen Hawking, A Brief History of Time, Chapter 6: "A cannonball fired upward from the earth will be slowed down by gravity and will eventually stop and fall back; a photon, however, must continue upward at a constant speed..."

FOOS: Hawking had it halfway right. If only he'd given it more thought. So light speed is not slowed as it travels upward as Einstein described in GR and as commonly accepted today? I used to think Hawking was an Einstein lackey, but now I'm thinking he was okay.

http://www.amazon.com/Why-Does-mc2-Should-Care/dp/0306817586

Brian Cox, Jeff Forshaw, p. 236: "If the light falls in strict accord with the principle of equivalence, then, as it falls, its energy should increase by exactly the same fraction that it increases for any other thing we could imagine dropping. We need to know what happens to the light as it gains energy. In other words, what can Pound and Rebka expect to see at the bottom of their laboratory when the dropped light arrives? There is only one way for the light to increase its energy. We know that it cannot speed up, because it is already traveling at the universal speed limit, but it can increase its frequency."

FOOS: Blue light does have more energy then red at a given gravitational potential, but light at a lower elevation is not at the same gravitational potential. Any gain in energy is offset by lost kinetic energy through slowing down, not speeding up. Proof of that is that Essen's invariant measurement of light speed and frequency throughout a gravitational gradient. Light neither changes speed or frequency throughout a gravitational potential.

http://briankoberlein.com/2014/08/19/red/

Brian Koberlein: "When we shine the flashlight upward, Newtonian gravity would say that the light is unaffected since light is massless, but under general relativity light is affected by gravity, so as the light travels upward it must lose energy. But how is that possible if it can't slow down?"

FOOS: Great point, but Newtonian gravity does not say that. Newton claimed that in a gravitational field the speed of light varies like ordinary bodies, the same as Einstein's GR model. But the measure of c by means of Essen's experimental value demonstrates c is unchanged, so both models are wrong.

http://galileo.phys.virginia.edu/classes/252/general_relativity.html

Michael Fowler, University of Virginia: "What happens if we shine the pulse of light vertically down inside a freely falling elevator, from a laser in the center of the ceiling to a point in the center of the floor? Let us suppose the flash of light leaves the ceiling at the instant the elevator is released into free fall. If the elevator has height h, it takes time h/c to reach the floor. This means the floor is moving downwards at speed gh/c when the light hits. Question: Will an observer on the floor of the elevator see the light as Doppler shifted? The answer has to be no, because inside the elevator, by the Equivalence Principle, conditions are identical to those in an inertial frame with no fields present. There is nothing to change the frequency of the light.

This implies, however, that to an outside observer, stationary in the earth's gravitational field, the frequency of the light will change. This is because he will agree with the elevator observer on what was the initial frequency f of the light as it left the laser in the ceiling (the elevator was at rest relative to the earth at that moment), so if the elevator operator maintains the light had the same frequency f as it hit the elevator floor, which is moving at gh/c relative to the earth at that instant, the earth observer will say the light has frequency $f(1+v/c) = f(1+gh/c^2)$..."

FOOS: CORRECT! Finally, a scientist who can see straight but fails to realize that the two different observations require the length of the local meter to be shorter by the reckoning of the Earth bound remote observer. This is Einstein's time dilation (original point of the Pound-Rebka experiment) which is synonymous with Einstein's length dilation by definition. Proof is the equivalent time delay (slowing of velocity) observed as Mercury approaches the sun; therefore, the speed of light observed remotely slows on approach (and increases after passing the sun) while no change is measured along the path.

In this example, the elevator (local) observer confirms no change in c. The outside (remote) observer confirms the frequency has increased (and hence c is slowed). The time dilation effects is equivalent to length contraction relative to the top even though the local observer sees no change. This means the velocity of light slows on approach with the contraction of length, not increases. The change is accounted for by reversing the sign; hence, $c(1-gh/c^2)$, not $c(1+gh/c^2)$. But this effect is seen by the Earth bound observer and not locally along the path of light because wavelength w and clock speed r are both reduced. Actually, the two are synonymous. . The Earth bound observer reckons w' and $r' = w$; $r(1-gh/c^2)$ as correctly documented in Figures 1 and 2:

Since the speed of light is distance per unit time, the downward speed of light c' is diminished by the reckoning of the "remote", or Earth based observer, because by definition space has contracted by the same amount. A meter is invariant wherever locally measured, but gravitational change in frequency requires by definition that $c' \equiv c(1-gh/c^2)$, the opposite of falling bodies $c(1+gh/c^2)$. Newton's corpuscular (emission) theory was wrong. Light has no mass and accelerates upwards opposite the rate of falling bodies if the definition and properties of a meter are to be preserved. c must be locally invariant as established by Essen, but remote observations of c, length and wavelength all increase together with elevation. This is what the 1960 Pound-Rebka experiment proved, not that the path of light is the same as bodies of mass.

To remove all doubts, the motion of Mercury when approaching the sun is delayed. Call it time dilation, be my guest, but the observer sees the velocity of light as being slowed by increased gravitational potential by $(1-gh/c^2)$; not increased by $(1+gh/c^2)$. Light accelerates away, not toward mass.

Valev: Einstein's increase in wavelength **(being $(1+2gh/c^2)$** is obviously absurd which means that the Pound-Rebka experiment has actually confirmed Newton and refuted Einstein. The below references show that, according to Einstein's general relativity, in a gravitational field the speed of light varies in conformity with the equation $c' \equiv c(1+2gh/c^2)$:

FOOS: Okay, okay, I get it, backwards, but that was a long time ago.

http://arxiv.org/ftp/arxiv/papers/1111/1111.6986.pdf

J.D. Franson, Physics Department, University of Maryland: "According to general relativity, the speed of light c as measured in a global reference frame is given by $c=c0(1+2phi/c0^2)$, where c0 is the speed of light as measured in a local freely-falling reference frame."

http://arxiv.org/pdf/gr-qc/9909014v1.pdf

Steve Carlip: "It is well known that the deflection of light is **twice that predicted by Newtonian theory**; in this sense, at least, light falls with twice the acceleration of ordinary "slow" matter."

http://www.speed-light.info/speed_of_light_variable.htm

"Einstein wrote this paper in 1911 in German. (...) ...you will find in section 3 of that paper Einstein's derivation of the variable speed of light in a gravitational potential, eqn (3). The result is: **c'=c0(1+phi/c^2)** where phi is the gravitational potential relative to the point where the speed of light c0 is measured. (...) You can find a more sophisticated derivation later by Einstein (1955) from the full theory of general relativity in the weak field approximation. (...) Namely the 1955 approximation shows a variation in km/sec twice as much as first predicted in 1911."

FOOS: Hair yesterday, gone tomorrow, back again. Now you see it, now you don't... But the perihelion of Mercury example requires the sign be reversed. c is observed to diminish on approach

http://www.ita.uni-heidelberg.de/research/bartelmann/Publications/Proceedings/JeruLect.pdf

LECTURES ON GRAVITATIONAL LENSING, RAMESH NARAYAN AND MATTHIAS BARTELMANN, p. 3: " The effect of spacetime curvature on the light paths can then be expressed in terms of an effective index of refraction n, which is given by (e.g. Schneider et al. 1992): **n = 1-(2/c^2)phi = 1+(2/c^2)|phi|** Note that the Newtonian potential is negative if it is defined such that it approaches zero at infinity. As in normal geometrical optics, a refractive index n1 implies that light travels slower than in free vacuum. Thus, the effective speed of a ray of light in a gravitational field is: v = c/n ~ c-(2/c)|phi|".

FOOS: More than a decade old. Here we see that the original fraudulent claim of Einstein's 2/c^2 was not yet amended to 1/c^2 in 1992. The square appears to be missing in the last expression.

http://www.mathpages.com/rr/s6-01/6-01.htm

"Specifically, Einstein wrote in 1911 that the speed of light at a place with the gravitational potential phi would be **c(1+phi/c^2)**, where c is the nominal speed of light in the absence of gravity... However, this formula for the speed of light (not to mention this whole approach to gravity) turned out to be incorrect, as Einstein realized during the years leading up to 1915 and the completion of the general theory. We now have c_r =1+2phi, which corresponds to Einstein's 1911 equation, except that we have a factor of 2 instead of 1 on the potential term."

FOOS: But now it's back to the original. Obviously, this was before the Pound-Rebka result. The articles on mathpages are undated, anonymous and crammed with pro Einstein gibberish.

http://poincare.matf.bg.ac.rs/~rviktor/kosmologija/Relativity_Gravitation_and_Cosmology.pdf

Relativity, Gravitation, and Cosmology, T. Cheng. p. 49: The speed of light as measured by the remote observer is reduced by gravity as c(r) = (1 + phi(r)/c^2)c. Namely, the speed of light will be seen by an observer (with his coordinate clock) to vary from position to position as the gravitational potential varies from position to position. Namely, the retardation of a light signal is twice as large as that given in (3.39), now c(r) = (1 + 2phi(r)/c^2)c.

Valev: The Pound-Rebka experiment has actually confirmed the prediction of Newton's emission theory that, in a gravitational field, the speed of light varies like the speed of ordinary falling objects (same acceleration), and refuted the prediction of Einstein's relativity that it varies twice as fast as the speed of ordinary falling objects (if, for light, the emission theory predicts acceleration g, general relativity predicts acceleration 2g.

http://www.einstein-online.info/spotlights/redshift_white_dwarfs

Albert Einstein Institute: "One of the three classical tests for general relativity is the gravitational redshift of light or other forms of electromagnetic radiation. However, in contrast to the other two tests - the gravitational deflection of light and the relativistic perihelion shift -, you do not need general relativity to derive the correct prediction for the gravitational redshift. A combination of Newtonian gravity, a particle theory of light, and the weak equivalence principle (gravitating mass equals inertial mass) suffices. (...) The gravitational redshift was first measured on earth in 1960-65 by Pound, Rebka, and Snider at Harvard University..."

Valev: Today's Einsteinians contradict Einstein. He taught the speed of light in a gravitational field was variable, they now teach it is constant:

FOOS: Backwards, isn't it? Anyway, it is locally constant.

http://bartleby.net/173/22.html

Albert Einstein: "In the second place our result shows that, according to the general theory of relativity, the law of the constancy of the velocity of light in vacuo, which constitutes one of the two fundamental assumptions in the special theory of relativity and to which we have already frequently referred, cannot claim any unlimited validity. A curvature of rays of light can only take place when the velocity of propagation of light varies with position."

FOOS: So, he talks out of both sides of his mouth. Nothing new here. What he means is that the local measure of c remains valid, but the remote measure of c exemplified in the Mercury orbit slows on approach to a body of mass, in which case the velocity of light in the Pound-Rebka experiment is $c(1-gh/c^2)$ instead of $c(1+gh/c^2)$. It can't be both ways.

http://math.ucr.edu/home/baez/physics/Relativity/SpeedOfLight/speed_of_light.html

Don Koks, Steve Carlip, Philip Gibbs: "So consider the question: "Can we say that light confined to the vicinity of the ceiling of this room is traveling faster than light confined to the vicinity of the floor?". For simplicity, let's take Earth as not rotating, because that complicates the question! The answer is then that (1) an observer stationed on the ceiling measures the light on the ceiling to be traveling with speed c, (2) an observer stationed on the floor measures the light on the floor to be traveling at c... Einstein talked about the speed of light changing in his new theory. In his 1920 book "Relativity: the special and general theory" he wrote: "... according to the general theory of relativity, the law of the constancy of the velocity of light in vacuo, which constitutes one of the two fundamental assumptions in the special theory of relativity [...] cannot claim any unlimited validity. A curvature of rays of light can only take place when the velocity [Einstein means speed here] of propagation of light varies with position." This difference in speeds is precisely that referred to above by ceiling and floor observers."

FOOS: So, light is c as measured both top and bottom according to Einstein's SR is constant, but with GR it is slower at the bottom? The curvature he refers to stems from the Eddington measurement which failed to account for atmospheric refraction. There is no curvature.

http://www.amazon.com/Why-Does-mc2-Should-Care/dp/0306817586

Brian Cox, Jeff Forshaw, p. 236: "If the light falls in strict accord with the principle of equivalence, then, as it falls, its energy should increase by exactly the same fraction that it increases for any other thing we could imagine dropping. We need to know what happens to the light as it gains energy. In other words, what can Pound and Rebka expect to see at the bottom of their laboratory when the dropped light arrives? There is only one way for the light to increase its energy. We know that it cannot speed up, because it is already traveling at the universal speed limit, but it can increase its frequency."

Valev: A desperately lying Einsteinian (who has not yet left the sinking ship):

FOOS: Yes, but a frequency change is not equivalent to a change in velocity.

http://briankoberlein.com/2014/08/19/red/

Brian Koberlein: "When we shine the flashlight upward, Newtonian gravity would say that the light is unaffected, since light is massless, but under general relativity light is affected by gravity, so as the light travels upward it must lose energy. But how is that possible if it can't slow down?"

Valev: Newtonian gravity would not say so, Brian Koberlein. It says that, in a gravitational field, the speed of light varies like the speed of ordinary bodies - this is confirmed by the Pound-Rebka experiment.

FOOS: An awful lot of confusion out there. Just because red light has less energy at a given gravitational potential does not mean it loses energy. The potential energy increases as the light moves upward at the same rate it loses energy according to Plank's hv and neither the frequency or velocity of the flashlight is changed along its path or Essen's c would be gravitationally dependent, which it is not.

Differences Between The Einstein and Ether Entrainement Theories: How To Trap Them Einsteins In A Black Hole

Herbert Dingle, President of the Royal Astronomical Society, tried it. He announced to the NY Times he had solid proof Einstein's theory was bunk. The Times did publish Dingle's proof, but purely rational as it was, it was denounced and ridiculed by scientists everywhere. Others tried and risked losing their careers. The game is rigged. The Big Bang Boozle is a straightforward proof of Einstein's fraud and correct solution to the cause of gravity. But there is no easy way to trap these little Einsteins, slippery snakes with their heads in the government trough. Running the gauntlet of vulgar smears in the Google group sci.physics.relativity shows 99% of the field consists of two main groups, the Einsteins and the ether entrainment theorists we will refer to as the Entrainers. Both are dug in deep for the Newtonian/Einstein model that light accelerates downward like ordinary objects, supported by their backwards interpretation of Pound-Rebka, the initial premise of The Big Bang Boozle. Einstein's special relativity assumes a constant c, but gravitational effects are general relativity and cite the Pound-Rebka experiment as proof c consists of photons and varies like objects of mass.

If we could put an Einstein or Entrainer on the witness stand and let justice be decided only by the immutable rules of ninth grade algebra, the procedure is quick and deadly. It also proves Newton could not have been the one who discovered the laws of gravitation. Who then? I dare not say, but it couldn't have been Newton since he didn't understand the full meaning of the universal law, $G=R^2/M * (c/s)$. We can firmly refute the claim that photons fall at rate g or any other rate. We can establish the real meaning of Newton's G and cause of gravity. Since Newton missed those, he could not have been the original author of the universal law of gravitation. A 9th grade algebra textbook problem overturns the laws of physics as now understood.

Laurence: "Light going up is redshifted because it is slowed by gravity." Valev: "Newtonian gravity says that, in a gravitational field, the speed of light varies like the speed of ordinary bodies - this is confirmed by the Pound-Rebka experiment."

To be sure, they mean velocity c really is variable and not just an inference. This consensus (shared by Einsteins) is well illustrated by Pentcho Valev, a leading Entrainer. Generally, the downward increase in light speed is $c'=c(1+gh/c^2)$ (Compare to Foos derivations in Figures 1 and 2). Valev cites a sufficient number of scientific references to establish consensus on this upside down interpretation: https://bit.ly/3yX1jXf) This (or the 2x version) is the common view of Einstein's claim and also identical to Newton's original corpuscular theory of the 1600s. One wonders how Einstein weaseled in, but the tone of Valev's essay is to declare Einstein a fraud because of the outdated, trivial claim that photons fall twice as fast as golf balls stemming from his 2x starlight deflection fraud. We need only care that both theories model photons as affected by gravity the same as other objects. The scientific community believes photons are pulled downward like any object at the rate $c'=c(1+gh/c^2)$. Nobody takes seriously Einstein's early claim that photons fall twice that fast. Hawking's invariant c is an exception, but that still leaves him in a black hole, not just dug in so deep. Explained later.

If you've grasped the consensus of quotations by Valev that photons fall at rate $(1+gh/c^2)$, then can everyone agree with the established scientific fact that G is the geometrically derived, immutable gravitational constant equal to $R^2/M*g$ everywhere in the universe for every body of mass of any size? Then do we agree we can scale this to R as radius of the universe and M mass of the universe and g is then equal to c/s? The value of g for the entire universe must be c/s. I we can't agree on that, we must stop and declare a mistrial.

So, having made it this far, can we agree that G holds equal to GM/R^2*g for a universe of any size? If not, we declare a mistrial. This can be verified by letting the sun be a universe unto itself and calculate that g, the rate of falling bodies, is identical to the 274 meters/s^2 in physics textbooks. Be honest. G is a geometrically derived constant and equal to $G=R^2/M*g$ for any universe, hypothetical or real. And can we also agree that since $G= R^2/M*c/s$, 274 meters/s is the speed of light in our little sun universe as we found it to be our own universe? This is obvious, so if we can't agree, it's a mistrial. The speed of light for a sun-is-everything universe is 274 m/s, the same as the calculated rate of falling bodes at g on the surface of the sun. This proves G is valid for any size universe and that c is equal to g at the surface of a universe. The kinetic energy of falling bodies at the surface is thus $E=mc^2$. These are all facts validated by 9th grade algebra. If we can't agree on them, the judge must grant a divorce. Are we still on the same page? So, in our sun universe, they all agree the Pound-Rebka results prove that photons fall downward at the same rate that all other objects do, ignoring Einstein's outdated claim they fall at twice that. Numerous quotes by Valev support that interpretation. This matches Laurence's statement above. Why not? That means the initial velocity of photons in our sun universe c is 274 m/s being shot out of little atomic cannons, and the rate they fall backwards is 274 m/s^2.

Do You see the problem, Mr. Columbo? The Einsteins, et al, are satisfied that photons are shot out of little cannons at speed c as Einstein described, and also believe the Pound-Rebka experiment proves photons also fall downward at the accelerating rate $c1=c*(1+gh/c^2)$. Photons accelerate backwards faster than their initial velocity c? What does the judge have to say? The escape velocity of c is less than the accelerating travel backwards. **So the sun can not shine! Let the judge decide.** As for Hawking, his invariant c over distance h matches the rigorous, established, universally accepted measurements of Louis Essen and so is safer than the Einstein/Entrainer model, but it means downward velocity c is equal to the initial escape velocity c, so photons are frozen in position instead of forced downward. This equals Hawking's own definition of a black hole. And that is how we trap all them little Einsteins in a black hole.

www.ingramcontent.com/pod-product-compliance
Lightning Source LLC
Chambersburg PA
CBHW051920210526
45473CB00006B/2085